To Derek

Thanks for your amazing co
to read hardly challenges

All best ~

MCKAL

After 30+ years since meeting you
in Scottish Homes ...
    All the best, ?Ay

.

RURAL POVERTY TODAY

Experiences of Social Exclusion in Rural Britain

Mark Shucksmith, Jayne Glass, Polly Chapman
and Jane Atterton

With a foreword by
Professor Sir Howard Newby

First published in Great Britain in 2023 by

Policy Press, an imprint of
Bristol University Press
University of Bristol
1–9 Old Park Hill
Bristol
BS2 8BB
UK
t: +44 (0)117 374 6645
e: bup-info@bristol.ac.uk

Details of international sales and distribution partners are available at
policy.bristoluniversitypress.co.uk

British Library Cataloguing in Publication Data
A catalogue record for this book is available from the British Library

ISBN 978-1-4473-6712-3 hardcover
ISBN 978-1-4473-6713-0 ePub
ISBN 978-1-4473-6714-7 ePdf

Cover design: Robin Hawes
Front cover image: iStock/Philip Silverman
Bristol University Press and Policy Press use environmentally responsible
print partners.
Printed in Great Britain by CPI Group (UK) Ltd, Croydon, CR0 4YY

Contents

List of figures and tables

Glossary

Deprivation Deprivation is a multidimensional, if contested, term. It gained currency during the 1970s when indicators of deprivation were used to target resources, leading councils in rural areas to point to rural deprivation as part of what became known as 'the arithmetic of woe'. Nevertheless, Shaw's study of rural deprivation (Shaw, 1979) was influential in conceiving three dimensions of household deprivation, opportunity deprivation and mobility deprivation. Sophisticated indicators of multiple deprivation, combining several domains, are still used to allocate resources in the UK, though these are often criticised for an apparent urban bias in their construction.

Financial hardship Financial hardship is understood by the Standard Life Foundation (SLF, now abrdn Financial Fairness Trust) as the counterpart to financial wellbeing, that is, a situation in which people are unable to meet financial commitments and needs comfortably. This might also be considered in relation to definitions of poverty used in the UK, namely, an inability to share in the lifestyles of the majority, or a household income below 60 per cent of the national median.

Financial vulnerability Financial vulnerability, according to the Financial Conduct Authority (FCA), refers to those adults who may suffer disproportionately if things go wrong because they have low financial resilience. It also covers those who may be less able to engage with their finances or financial services. The reasons for this may vary from experiencing a particular life event (such as a bereavement or redundancy) to having low financial capability. 'In difficulty' refers to adults who are the least financially resilient, as they have already missed paying bills or meeting credit commitments in at least three of the last six months. FCA analysis in 2018 found that 54 per cent of all rural consumers (sic) in the UK were financially vulnerable, defined as having one or more of these characteristics.

Financial wellbeing For the SLF, this term refers to improving people's ability to meet financial commitments and needs comfortably and a situation whereby an individual has enough income for more than life's essentials and is not struggling to make ends meet. It means having the capacity to do this in future, including the ability to deal with financial shocks and to have saved enough for retirement.

Poverty Poverty can be understood in a narrower or a broader sense, and in relative or absolute terms (see Shucksmith and Schafft, 2012). Following Townsend (1979; 1987), the European Union (EU) conceives of poverty existing when people's resources (material, cultural and social) are insufficient to allow them to enjoy a minimum acceptable way of life. They are often excluded and marginalised from participating in activities that are the norm for other people and their access to fundamental rights may be restricted (Commission of the European Communities, 1993). From a statistical point of view, in Europe, people are regarded as relatively poor if their household income is less than 60 per cent of the median household income nationwide (Vera-Toscano et al, 2020).

Social exclusion Social exclusion is a dynamic, multidimensional concept, referring to the processes operating at various scales which allocate resources in societies, and through which advantage and disadvantage are conferred (Room, 1994). Philip and Shucksmith (2003) argue that these processes include markets (not only labour markets but housing, finance and so on), state (bureaucratic), community (associative) and friends and family (reciprocal). These also constitute the opportunity structures in which capacity, agency and agility may be built, supported and exercised. Social exclusion emphasises the relational aspects of life – social, cultural, political – as well as the distributive or material (Bailey et al, 2004).

About the authors

Mark Shucksmith OBE is Professor of Planning at Newcastle University. His research interests include social exclusion in rural areas and rural development. Recent books include: *Hope under Neoliberal Austerity: Responses from Civil Society and Civic Universities* (Policy Press, 2021); *Routledge International Handbook of Rural Studies* (Routledge, 2016); and *Rural Transformations and Rural Policies in the UK and US* (Routledge, 2012). Mark chaired the Scottish government's Committee of Inquiry into Crofting, and he remains a Trustee of the Carnegie UK Trust, and of Action with Communities in Rural England. Mark was awarded an OBE in 2009 for services to rural development and to crofting.

Jayne Glass was Research Fellow in the Rural Policy Centre at Scotland's Rural College (SRUC) at the time this research was carried out. She is now Researcher at Uppsala University in Sweden and remains an Honorary Lecturer at the University of Edinburgh. Jayne's research and teaching topics include community empowerment, rural community resilience (with a focus on children and young people), land reform and rural land use policy. She has published widely on these topics, often working closely with communities and other stakeholders to understand people's lived experiences of contemporary issues in rural areas. She recently co-authored *Land Reform in Scotland: History, Law and Policy* (Edinburgh University Press, 2020) and supported the Council of Europe's Congress of Local and Regional Authorities in their work on the future of youth in rural areas (2021–22).

Polly Chapman is the CEO of HISEZ CIC, a social enterprise based in Inverness which owns and operates Scotland's first Impact Hub, Impact Hub Inverness, serving the Highlands and Islands. As well as providing a co-working and meeting space, Impact Hub Inverness also provides business support to social enterprises across the Highlands and Islands. Polly has worked in the broad area of rural and community development for over 20 years, and 25 years ago worked with Mark Shucksmith at the University of Aberdeen on various research projects, including a book *Rural Scotland Today: The best of both worlds?* (1996) and several journal articles.

Jane Atterton is Senior Lecturer and Manager of the Rural Policy Centre at SRUC. She has over 20 years' experience researching rural issues, working in both academic and policy environments, including at Aberdeen and Newcastle Universities and the Countryside Agency. Jane's research interests include rural community change, including demographic shifts and processes of inclusion/exclusion; rural economies and enterprises; and rural policies and the policy-making process, including rural proofing and place-based policies. Jane's current role combines research and knowledge exchange activities, including supporting the Cross-Party Group in the Scottish Parliament on Rural Policy.

Acknowledgements

The authors are indebted to all those who participated in this research, whether through offering us advice and information, or by agreeing to be interviewed. Your names are not listed anywhere in the book to preserve confidentiality, but we are immensely grateful: we hope we have done justice to your stories.

We also wish to thank the members of our distinguished Advisory Group for their comments, advice and support throughout the project. Members of the Group included: Margaret Clark, Nicola Crook, Derek Egan, Tim Goodship, Rebecca Graham, Vanessa Halhead, Linda Hutton, Sarah Kidd, Karen MacNee, Coinneach Morrison, Angus Murray, Anne Murray, Michael Nixon, Norma Robson, Carol Tannahill and Ellie Thompson. Colleagues within our own organisations, Newcastle University, SRUC and Impact Hub Inverness, have also supported this project in many ways.

We are also grateful to those who spoke at our two online launch webinars in March 2021, organised in conjunction with a Member of the Scottish Parliament (John Scott MSP) and with the Rural Coalition respectively. In Scotland speakers included Aileen Campbell MSP (Cabinet Secretary for Communities and Local Government), John Scott MSP, Rhoda Grant MSP, Jamie Halcro-Johnston MSP, Dame Barbara Kelly, Vanessa Halhead (Scottish Rural Action), Calum Macleod (Community Land Scotland), Douglas White (Poverty & Inequality Commission) and Linda Hutton (Citizens Advice Scotland). Speakers at the England event included the Bishop of St Albans (Chair of Rural Coalition), Graham Biggs (Rural Services Network), Jeremy Leggett (ACRE), James Alcock (Plunkett Foundation), Claire Maxim (Arthur Rank Centre), Ellie Thompson (Trussell Trust) and Mubin Haq (Chief Executive of Standard Life Foundation [SLF]).

Finally, this research would not have been possible without the generous financial support of the SLF (now abrdn Financial Fairness Trust) and the valuable guidance of its officers, Rebecca Graham and Charlotte Morris. The SLF supported this as part of its mission to contribute towards strategic change which improves financial wellbeing in the UK. The Foundation funds research, policy work and campaigning activities to tackle financial problems and improve living standards for people on low-to-middle incomes in the UK. It is an independent charitable foundation registered in Scotland (SC040877). We also received financial support from Research Councils UK's Strategic Priorities Fund.

Foreword

Professor Sir Howard Newby

John Constable has a lot to answer for. The celebrated landscape artist was responsible for the introduction of the picturesque from continental Europe in the early 19th century, a convention that still dominates perceptions of the English countryside to this day. Most visitors to the countryside expect to find something 'as pretty as a picture' and that picture is epitomised by Constable's paintings. The most celebrated is *The Hay Wain*, in which a sturdy yeoman is seen driving his cart across the River Stour, in Dedham Vale on the Essex/Suffolk border. However, all is not what it seems.

Constable's portrayal is not an accurate one, nor was it intended to be. It is an elegy to a lost past, which if it existed at all had long disappeared by the 1820s when the painting was conceived. The open common land across the river had been subjected to enclosure and the independent yeoman would by now have been a destitute farm labourer casually employed, if employed at all.

Conditions were so bad that they provoked extensive social unrest – the Captain Swing riots – with widespread arson, the maiming of animals and destruction of crops. East Anglia was where the rioting was most commonplace and where the authorities cracked down hardest. A picturesque rural idyll it was not.

As the 19th century proceeded poverty became increasingly viewed as an urban problem as the new industrial centres grew apace and the living conditions of the poor became only too obvious. In the countryside poverty remained mostly unacknowledged. For the rural poor Methodism provided some solace, but trade union organisation largely failed, notwithstanding the efforts led by Joseph Arch in the 1870s. Poverty still remained hidden. The poor made little fuss and slipped away to better-paid employment when they could. There was often a view that they were poor but happy; whereas the urban poor were poor and miserable.

Although there was a considerable exodus from rural areas of able men to fight in the First World War, low farm wages continued to prevail, so that this remained the economic basis of rural poverty. It took another World War to produce an extraordinary transformation in the nature of the rural economy and the demographic characteristics of the rural population.

In the long arc of this process from the late 1940s to the mid-1970s both the rural economy and the rural population irrevocably changed. And the nature of rural poverty therefore changed too.

The first driver of this change has often been called the Second Agricultural Revolution, typified by the shift from a horse-and-hand technology to one based on the internal combustion engine (though there was much more to it than this). The demand for farm labour plummeted and even those farm workers who remained found that their newly acquired skills could be deployed more lucratively elsewhere – in the emerging road haulage industry, for example.

As farm workers left the villages they were replaced by a new kind of rural dweller: professional and managerial urban residents who took advantage of the opportunity afforded by the internal combustion engine to work in the towns and cities, but live in the countryside. They were attracted by cheaper housing (until the 1960s when this price advantage disappeared), to escape from the manifest problems of urban life and to embrace precisely that picturesque vision of life in the countryside which continued to prevail in the minds of those whose only encounter with the countryside was occasionally to visit it. Two nations in one village was often the outcome.

So, the second regulator of rural change became the nature of the rural housing market. Demand increasingly outstripped supply as commuters, second homeowners, retired couples and holiday landlords comfortably outbid the locals for house purchase and as holiday lets reduced the pool of rental property. Right to buy legislation in the 1980s also shrank the amount of social housing as these properties were improved and then sold for a considerable profit. Affordable homes often disappeared completely from many communities and could not be replaced due to the stern resistance of the ex-urban population, for whom the appearance of the countryside was paramount and new housing regarded as an intrusion.

This has produced a situation in which the entire social and cultural character of village communities has changed. Villages look more visually attractive than at any time in their history, but with the change in the population many of the institutions of village life have withered away: schools, pubs, churches, post offices and shops. They were followed by the decline of rural services – transport, health and so on – and the non-existence of others, most recently the access to broadband communications.

So, the sum total of all of this is that the nature of rural poverty has changed. With agricultural employment accounting for less than 5 per cent of the rural labour force, farm wages can no longer be regarded as the major cause of rural poverty (which is not to say that farm wages are high – they are not – but they are not by themselves sufficient to understand the contemporary distribution of the rural poor).

The modern rural economy is overwhelmingly a service economy and primary in this is tourism – or hospitality in the cosy modern idiom. Unfortunately, this is also a source of poor pay and conditions. The myriad of cleaners, shop assistants, catering staff, bar staff and so on are beset by low pay, split shifts and zero-hour contracts. Rural housing is beyond their reach, although some, in a modern echo of agricultural tied cottages, have accommodation supplied by their employer. Many live in nearby towns, taking advantage of the larger pool of rented accommodation, and commute out to the countryside. Rural poverty can now be exported to nearby towns, making our understanding of the phenomenon more complicated.

A completely arbitrary example is Evesham in Worcestershire, a once-prosperous market town that has fallen on hard times. The modern shopping mall is almost empty, high street shops are boarded up and among those that remain, takeaways, charity shops and nail bars predominate. And yet just up the road there is Broadway, an archetypal Cotswold village and a honeypot for visitors. Is this a coincidence?

Some things have not changed. The rural poor are mostly scattered in small numbers, are not organised and so remain unnoticed and unreported. They do not disturb the idyllic view of rural life which has prevailed since Constable was painting 200 years ago. If levelling up is to be at all meaningful it must apply not only to north and south, but to the poor within and between rural communities. This makes this new study of *Rural Poverty Today* even more timely and relevant.

Introduction

In Britain, unlike in many other countries, there is a general impression that **poverty** is a largely urban phenomenon, associated with inner cities, difficult to let housing estates or ex-industrial areas. Rural areas are imagined as picturesque, idyllic and far removed from such hardship and **deprivation**. Perhaps for this reason, poverty and vulnerability in rural Britain are neglected both by research and by policy and practice. Yet analysis of the British Household Panel Survey shows 50 per cent of rural households experienced poverty at some time between 1991 and 2008, compared with 54 per cent in urban Britain (Vera-Toscano et al, 2020). Poverty is clearly a rural phenomenon too but, as Newby recognised (1980, 278):

> It is easy to overlook these problems amidst the general prosperity of contemporary rural England. The appearance of many villages suggests two-car families enjoying a lifestyle of comfortable affluence in their beautifully restored homes. The other face of rural England is more difficult to seek out since it is less openly admitted.

Addressing **financial hardship** among the 11.3 million inhabitants of rural Britain (ONS, 2016) is hampered by an inadequate evidence base, according to two House of Lords Select Committees, a Scottish Parliament Cross-Party Group and the Welsh Senedd. Analysis by the Financial Conduct Authority (FCA) (2018) revealed that 54 per cent of rural dwellers were financially vulnerable (compared with 48 per cent of urban residents) but not how or why. Levels of financial stress are known to be high, with fewer than half of those in the Highlands and Western Isles 'coping well or very well financially' (Scottish Household Survey, 2012), and there is widespread fuel poverty in rural areas (Scottish Government, 2018b; DEFRA, 2021). Despite this knowledge, research and policy focus overwhelmingly on urban experiences of hardship and 'austerity urbanism' (Peck, 2012).

These issues are important now because poverty and vulnerability have been increasing again, in all areas, exacerbated further by the effects of the COVID-19 pandemic. Alongside falling real incomes, rising debt, welfare reform, Universal Credit and Brexit uncertainties, pre-pandemic research revealed a marked acceleration in the loss of services and population decline in sparse areas (Wilson and Copus, 2018) and a more widespread loss of rural services in England (Rural England CIC, 2022). Some writers speculate that rural lives are better adapted to austerity, through years of underspending by the state and stronger voluntary and community effort, while others argue that rural areas are more vulnerable to withdrawal of those services that do remain, with no definitive evidence either way (Milbourne, 2016a). Indeed, does a high expectation of rural communities' social cohesion and potential for self-help legitimise and facilitate the dilution of state welfare systems and the withdrawal of the state from its responsibility to rural citizens? What are the implications for the UK government's policy objective of 'levelling up' and its shared responsibility to ensure wellbeing for all? Apart from a few studies (Milbourne, 2016b; Black et al, 2019; May et al, 2020), the effects of these structural processes on the lives and opportunities of citizens in rural Britain are unresearched, unlike in the United States (Sherman, 2009; 2021; Tickamyer et al, 2017) and Europe (Bernard, 2019; Bernard et al, 2019). Our work seeks to fill that gap and so to inform understandings and expectations of 'rural citizenship' (see, for example, Painter and Philo, 1995; Yarwood, 2017).

This book presents and reflects on findings from new research on why and how people in rural areas experience and negotiate poverty and **social exclusion**. It examines the roles of societal processes, individual circumstances and various sources of support (markets, state, voluntary and community organisations and family and friends).

Our conceptual framework focuses on the *interconnections* between individuals' and households' experiences and the structural and external processes bringing changes, for example, in local economies, employment, housing markets, welfare support and services. This is pursued through analysis at both the individual/household level, enquiring about experiences and causes of financial hardship and vulnerability and revealing coping strategies and sources of support, alongside analysis of the economic, social and policy context through which processes of social exclusion operate to generate or redistribute risk, vulnerability or hardship. These include narratives of place and belonging. Indeed, the analysis foregrounds the role of place as another dimension of intersectionality, examining how place modifies and intensifies the effects of other social characteristics such as class, gender and age. Previous scholars (Wyn and White, 2015) have argued, drawing on Bourdieu and Beck, that place constitutes social relations and they emphasised the importance of recognising belonging – the interconnections between biography and social context. We therefore pay attention to aspects of place identity and belonging as well as local opportunity structures (Bernard, 2019) as further significant factors in social exclusion/inclusion alongside standard dimensions of class, gender, age and ethnicity.

This conceptual framework connects our study's findings with wider sociological concepts of individualisation, risk society, precariatisation, labour market flexibilisation, welfare conditionality, 'roll-back' and 'roll-out' neoliberalisation and austerity urbanism, for example, so bridging between rural studies and wider sociology and social policy literatures. At the same time, the framework highlights emergent agency (individual and collective) and topical issues around civil society and community empowerment, along with current UK government policy narratives of loosely defined 'levelling up' (which includes policies to address regional inequalities and/or social inequalities) and 'building back better' (which suggests, but does not provide, a progressive vision).

In implementing this framework we have used financial hardship and vulnerability as a heuristic prism through which to uncover the broader processes of social exclusion. In part, this was because our funder, the Standard Life Foundation (SLF) (now renamed the abrdn Financial Fairness Trust) has relieving financial hardship and vulnerability as its charitable purpose. We were fortunate that the Trust allowed us to interpret this broadly to enquire into the underlying processes of social exclusion. Indeed, our funders were supportive and encouraging in every respect.

The next two sections of this introduction offer brief summaries of recent trends in poverty in Britain and of the changing policy context. These will be important in understanding changes observed in all our study areas. The research methods employed in this study are also briefly outlined.

Poverty in Britain: recent trends

The incidence and nature of poverty in Britain have changed during the last few decades in several respects, and these changes are evident in rural Britain too. Poverty has become less associated with unemployment and with old age: most low-income households are now in work, especially in rural areas; and rather than old age, now 'you are much more likely to be in poverty if you live in certain regions, live in a family where there's a disabled person or a carer, if you work in certain sectors such as accommodation and catering or retail, or if you live in privately rented housing' (Joseph Rowntree Foundation, 2020, 2).

Poverty in the UK, and in Europe generally, is usually defined in relative terms as a household income falling below two thirds of the country's median household income. The UK Department of Work and Pensions (DWP) collects information annually on poverty through its Households Below Average Income (HBAI) survey, and we can observe trends over the past 50 years from this, as shown in Figure 1.1. Poverty in Britain is shown both before housing costs (BHC) and after deduction of housing costs (AHC): the widening gap between these lines reflects a rapid rise in housing costs facing lower-income groups in Britain. Indeed, evidence is emerging that housing costs have been rising even more rapidly for lower-income groups in rural Britain who find themselves increasingly in private rented housing (Vera-Toscano et al, forthcoming).

Figure 1.1: Percentage of the British population in relative low income (1961–2018)

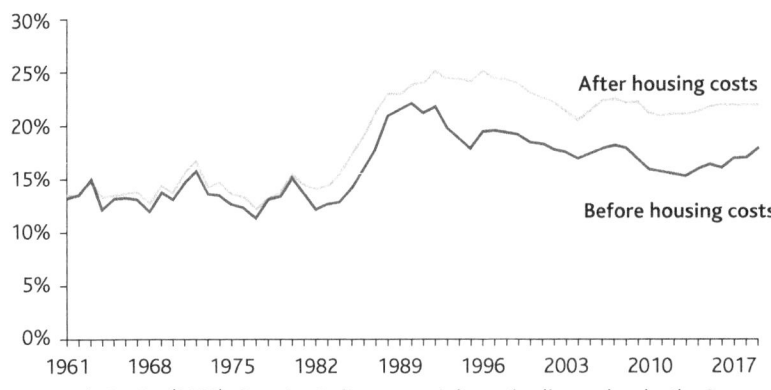

Source: Francis-Devine (2022). Contains Parliamentary information licensed under the Open Parliament Licence v3.0.

Broadly, poverty increased sharply in the 1980s during the Conservative governments led by Margaret Thatcher, stabilised under John Major's Conservative government and then fell under the Labour governments of Tony Blair and Gordon Brown. After 2012 poverty began to rise again under the Conservative-led coalition government. Figure 1.2 disaggregates these data (AHC) by household type for the last 25 years, showing a very marked decline in poverty in old age from 1997 to 2010, alongside smaller variations in poverty for other social groups.

Specifically, according to the HBAI data, poverty in old age fell from 24 per cent in 1997 to 14 per cent in 2010, then had increased slightly to 16 per cent by 2018. Poverty among working-age adults without children has been fairly stable at around 19 per cent since 2010, while for those with children it has been higher at around 25 per cent. Nevertheless, child poverty fell from 34 per cent to 27 per cent between 1997 and 2010 before increasing again to 31 per cent by 2019.

A further important feature of the last decades has been a steady rise in in-work poverty, 'because often people's pay, hours, or both, are not enough. Around 56 per cent of people in poverty are in a working family, compared with 39 per cent 20 years ago' (Joseph Rowntree Foundation, 2020). In 2009 an analysis of the DWP's HBAI dataset showed that 'poverty in work' was much more likely in rural than urban areas, and highest (67 per cent) in the most rural council areas, as shown in Figure 1.3).

The changing policy context

These trends in poverty in Britain are closely associated with policy, and specifically with the abandonment of the Keynesian economic

Figure 1.2: UK poverty rates for children, working-age adults and pensioners (1994–2020)

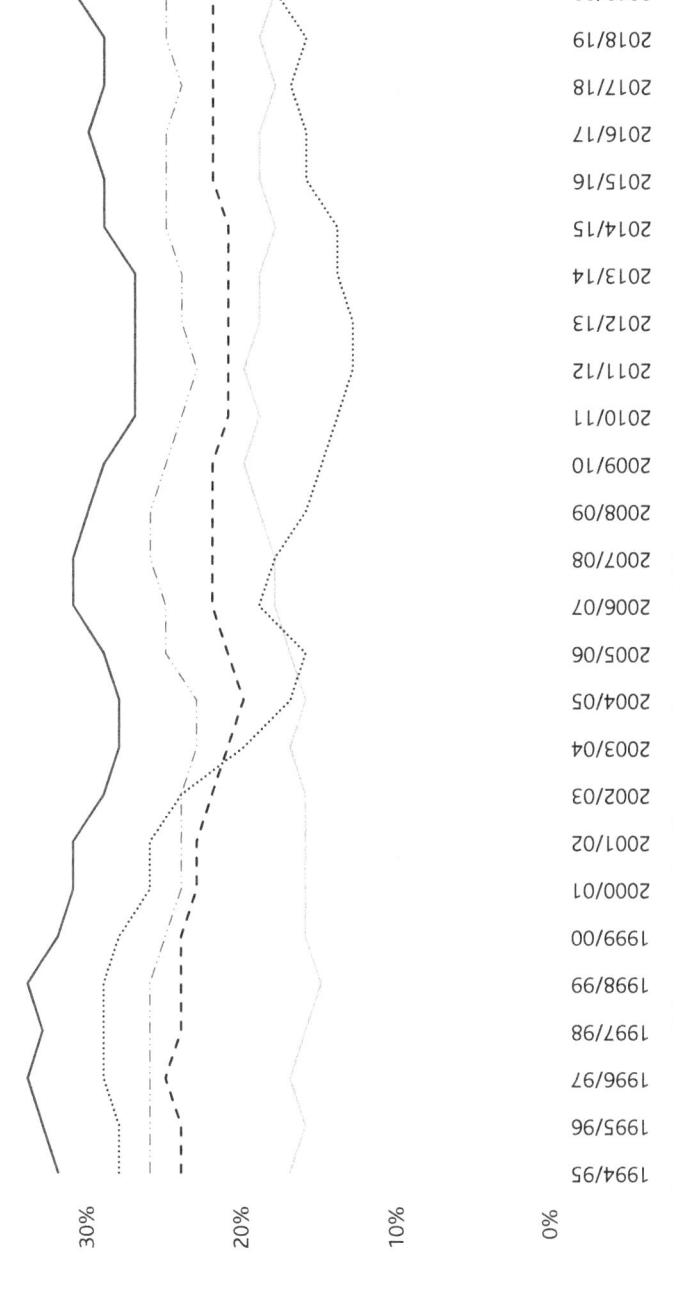

Source: Joseph Rowntree Foundation (2022) 'UK Poverty 2022: the essential guide to understanding poverty in the UK'

Figure 1.3: Share of UK working-age adults living in households with less than 60 per cent of median income 2005–08 (after housing costs)

Note: In the more rural districts around two thirds of working-age people with low incomes live in households where someone works. Only a third live in workless familes.
Source: Palmer (2009)

paradigm during the 1980s and its supplanting by neoliberalism, alongside globalisation. Neoliberalisation is often understood as having two phases. The 'roll-back' phase of neoliberalisation involved shrinking the state and the institutions of Keynesian welfarism and social collectivism through privatisation and cutting public expenditure, which were all features of the 1980s in Britain. The later, emergent 'roll-out' phase involved the purposeful construction and consolidation of neoliberalised state forms, modes of governance and re-regulation (Peck and Tickell, 2002). Both aspects of neoliberalisation had important impacts on rural economies and societies. Cuts to public expenditure (roll-back) reduced the funding available to support civil society institutions, while leaving ever larger gaps in state social provision for civil society to try and fill. Thus, healthcare, schools, council offices, churches, banks, pubs, libraries, shops, post offices, public transport as well as emergency services have been gradually lost. Meanwhile, neoliberalised economic management (roll-out) requires voluntary and community organisations to act competitively rather than collaboratively, to seek income from the state through tenders and contracts, to pursue targets and deliverables tangential to their purpose and in short to adopt the modes and practices of private business to pursue the agendas of the neoliberal state.

Prior to the 1980s, the post-war welfare state operated as a risk-pooling mechanism, underpinned by values of solidarity. However, the reconfiguration of welfare regimes 'towards an increasingly personalised responsibility for managing life's adversities' (Asenova et al, 2015, 14–15) has contributed to a modern 'risk society' (Beck, 1992; 2000) characterised by profound uncertainty and individualisation of responsibility and risk. The pace of change, the increased role of impersonal systems and institutions, notably global markets, and the rise of insecure employment means that the ability to survive and prosper has become more precarious for many. Indeed, in his book *The Precariat*, Standing (2011, 1) argues that a central theme of neoliberalism was that 'countries should increase labour market flexibility, which came to mean transferring risks and insecurity onto workers and their families', especially when accompanied by a welfare reform which creates insecurity also around social income and self-identity. During the 1990s, following the 'welfare to work' approach being deployed by the Clinton government in the US, the Conservative government made numerous reforms to social welfare policies in Britain to make it harder to claim, primarily by increasing welfare conditionality. One of these was the replacement of unemployment benefit and income support with Job Seekers' Allowance (JSA), which reduced benefit levels and increased conditionality by introducing more stringent job search requirements.

Labour governments from 1997 continued applying these conditions but offset them with measures to make work more rewarding and achievable, notably through a suite of New Deal schemes and the introduction of a National Minimum Wage (NMW) in 1999. Of most relevance here are Tax Credit, Pension Credit and Incapacity Benefit. Working Family Tax Credit was introduced as a work-contingent, means-tested benefit for parents with dependent children in 1999 and (renamed the Working Tax Credit) extended to all those of working age (with or without children) from 2003. These tax credits not only augmented family incomes but also included an allowance towards childcare costs. Tax credits, together with increases in Child Benefit rates, are generally regarded as effective in reducing child poverty.

A similar mechanism, Pension Credit, was introduced to address poverty among people of pensionable age, topping up their weekly income to a minimum level set by government. This also proved highly effective, so long as it was claimed, but analysis by Bradshaw and Richardson (Commission for Rural Communities, 2007) in England revealed that take-up is significantly lower in smaller rural settlements. Eligible residents not claiming Pension Credit were 35 per cent in urban areas, 43 per cent in villages and 54 per cent in hamlets and open countryside. Both these measures were administered not by DWP but by the tax authorities (HMRC), nor were they included in the definition of social assistance. Incapacity Benefit (supplanted in 2008 by Employment Support Allowance [ESA]) is also an important source of welfare support in rural areas. Labour's social policy reforms are generally agreed to have had a marked impact nationally in reducing poverty for low-income families and pensioners (Brewer et al, 2009; Cappellari and Jenkins, 2009; Joseph Rowntree Foundation, 2018), and analysis of British Household Panel Survey (BHPS) longitudinal panel data by Vera-Toscano et al (2020) concluded that these policy reforms also played an important role in reducing rural poverty over this period.

Following the financial crisis of 2007–08, the Labour government initially responded with a fiscal stimulus, but the election of Conservative-led governments from 2010 led to marked changes in policy in line with purer forms of neoliberalism. The imposition of public expenditure cuts and 'austerity' fell unevenly, with welfare spending and local government particularly hard hit.

The Conservatives' programme of welfare reform was intended to reduce public spending, reducing working-age welfare spending by £36 billion by 2019–20 (Keen, 2016), and to intensify work activisation. The reforms included reductions in working tax credits; a cap on how much benefit any one household could receive; a 'two-child limit' on Child Benefit; changes to the local housing allowance; the so-called 'bedroom tax' (which penalises benefit recipients in social housing with spare bedrooms); council tax benefit (now at the discretion of each council); and the introduction of Universal Credit (UC). UC was an attempt to streamline benefits, replacing several benefit streams (JSA, ESA, Income Support, Working Tax Credit, Child Tax Credit and Housing Benefit), but it has proved highly complex, can only be claimed online and requires a waiting period of several weeks during which no benefit is paid. Meanwhile, working-age benefits have not kept pace with living costs: from 2012, three years of below-inflation increases, followed by a four-year benefit freeze, eroded the value of many working-age benefits (Joseph Rowntree Foundation, 2021, 5). These measures, along with increasing sanctions and conditionality, added to the pressures on those with low incomes (Joseph Rowntree Foundation, 2021). A report by the National Audit Office (NAO) has been highly critical of these reforms and their impact on the most vulnerable (NAO, 2018). Particularly hard hit were lone parents, people with disabilities, those experiencing in-work poverty, families with children and young people (Asenova et al, 2015). It is notable that childless households were more vulnerable to changes in earnings after the recession but less vulnerable to changes in welfare policies, unlike pensioners or families with children who had gained from Labour's social policy reforms and were therefore more vulnerable to policy changes (Francis-Devine, 2021). Continuing support throughout this period for pensioners, who benefited from a triple lock ensuring their pensions grew in real terms, contrasted with these cuts made to working-age benefits.

Local government spending was hit even harder: from 2010 to 2015 local authorities in England lost 27 per cent of their spending power (Hastings et al, 2015) with further cuts of 56 per cent over the next five years announced in 2015 (Hastings et al, 2017). Analysis by the Institute for Fiscal Studies (IFS) (Francis-Devine, 2021) concludes that these cuts were severe but were uniformly applied across councils pre-2014 and, moreover, that this changed little following the introduction of a new formula for allocation from 2013 to 2014. This is disputed by May et al (2020) who reanalysed this data using a different rural definition. Hastings et al (2015, 6) also found that the cuts were uneven with poorer areas hit hardest. The 'austerity urbanism thesis' (Peck, 2012) posits these cuts have been focused on cities, and that within cities local authorities passed these on to the poorest in society. Notwithstanding this thesis, many rural councils have also experienced drastic cuts in central government funding and complain of lower funding per head than in cities. May et al's (2020) analysis, confirmed by Vera-Toscano et al (forthcoming), indicates that within rural Britain too, it was the most deprived areas which suffered the largest cuts in central government funding of local authorities.

By the onset of the COVID-19 pandemic the challenges of maintaining and delivering services in rural areas had already been heightened by a decade of these cuts to central government's contributions to English local authority budgets. The NAO (2021) found this contributed to a fall of around a third in councils' spending power which, alongside rising demand for services, had left councils more vulnerable to the impacts of the pandemic. The NAO warns of continuing cuts to services in the next few years, including social care, special educational needs, libraries, buses and community centres, as councils struggle to meet the extra costs incurred during the pandemic: 94 per cent of English councils expect to have to cut spending in 2021–22 to meet legal duties to balance their budgets, and several risk insolvency. This pressure is likely to lead to further centralisation or loss of public services in rural areas and provides further context to this study.

Much of this policy context also applies in Scotland, where welfare benefits, pensions and social security are 'reserved' matters still mostly under the control of the UK government (within DWP). Scotland's overall budget is also determined by the UK government through the 'Barnett formula' so that, in this respect too, austerity has been imposed on Scotland. However, the Scottish government has not subjected councils to the substantial cuts imposed on those in England. Moreover, some powers over social security have begun to be devolved to Scotland, following the Scotland Act 2016 and the Social Security (Scotland) Act 2018. At the time of our fieldwork, only minor powers had been devolved, but these will eventually include autonomy over disability and illness benefits, measures to address child poverty, variations in administration of UC and new benefits in areas of devolved responsibility or to top up reserved ones. Among the areas which remain reserved to the UK's DWP are UC, pensions, Pension Credit and Child Benefit. The Scottish government aims 'to create a Scottish social security system based on dignity, fairness and respect', for example, by replacing work capability assessments contracted out to the private sector with consultations with trusted health professionals, campaigning to improve benefit take-up and 'engaging with people with experience of receiving benefits, to build a social security system that works for them' (Scottish Government, 2021). In November 2021, after this research was completed, the Scottish government announced a doubling of the recently introduced Scottish Child Payment to £20 per week as part of a strategy to address child poverty. Scotland already has several other differences from England in relation to living costs, including free school meals for all children in classes Primary 1 to Primary 5, free prescriptions and free university tuition. There are also important differences in relation to social care provision. In 2016 the Scottish government announced that all publicly funded, frontline social care workers should be paid the real living wage. Furthermore, there are more care staff on permanent contracts in Scotland and personal care is funded for people at home, the means test is more generous and spend per head on social care is higher than in England. Nevertheless, care worker recruitment and retention remain difficult in Scotland.

There are also differences between England and Scotland in relation to rural policy. While neither country has a current rural policy statement, there has been a stronger commitment to rural issues in Scotland than in England in recent years. In England, the UK government has been under pressure from rural stakeholders and from a House of Lords Select Committee to adopt a rural strategy, but it rejects such an approach, relying instead on 'mainstreaming' through seeking to ensure that all mainstream policies are 'rural proofed' (considering whether policy is likely to have a different impact in rural areas, because of particular rural circumstances or needs; see Atterton, 2008). However, successive studies show that rural proofing is rarely undertaken and (in its present form) is ineffective (Shortall and Alston, 2016). Scotland developed innovative rural development policies in the 1990s and has subsequently introduced several new measures, such as community-based land reform and 'island proofing' (see Atterton, 2019), while giving much greater recognition to the needs of rural communities in the Scottish Parliament and in the post of a Cabinet Secretary for Rural Affairs and Islands. Somewhat paradoxically, therefore, Scotland's lack of an explicit rural policy and of rural proofing has been explained in terms of the *greater* importance of rural areas in Scotland's politics more generally (European Network for Rural Development, 2017, 32).

The commitment of the Welsh Assembly government to rural wellbeing is reflected in their funding of the Welsh Rural Observatory with a remit to support evidence-based rural policy in Wales. Despite this, Wales has no explicit rural policy statement either, relying on rural proofing instead. A report on rural poverty was prepared for the Welsh government by the Public Policy Institute for Wales (2016) and notes the evidence gaps. Furthermore, a 'Rural Health Plan – Improving Integrated Service Delivery Across Wales' (2009) sought to ensure that the future health needs of rural communities are met in ways which reflect the diverse conditions of rural Wales.

The context for this research, while varying in some respects between England and Scotland, is therefore one of fiscal tightening, labour market flexibilisation, welfare reform, neoliberalisation and the lack of an explicit, coherent rural strategy.

Methodology

The methods used in this study were subject to the normal ethical review procedures at Newcastle University, as well as critical review by the funders, SLF. The study began with a review of previous work on this subject, reported in Chapter 2, and an analysis of various secondary data sources to reveal national and local trends and to help contextualise our study areas in relation to national and regional data. These results are woven into subsequent chapters.

Fieldwork was carried out in three case study areas, chosen to reflect different types of rural area and circumstances, and our own familiarity with the areas from previous work. Two are in Scotland: an accessible rural area, East Perthshire, and a remote island area, Harris; and one in a remote mainland area in England, the North Tyne valley. Budgetary constraints prevented the addition of further study areas, but we were able to compare our results with recent work in Cornwall (Willett, 2021) and in Wales (Public Policy Institute for Wales, 2016), and to test the wider validity of our findings with stakeholders from across England and Scotland at two webinars. Members of England's Rural Coalition and other national and local stakeholders confirmed the relevance and validity of our findings to their locales. The locations of our three study areas are shown in Figure 1.4. The study area in East Perthshire includes the wards of Blairgowrie and the Glens and part of Strathmore. It has a population of about 19,000 and covers an area of 468 km^2. Half its residents live in Blairgowrie and Rattray, the principal town, while fertile lowlands including smaller towns stretch to the south, and a series of remote glens stretch north up into the Grampian mountains. The area is renowned for the growing of soft fruit and its rich past in textile weaving. Its landscape attracts tourists, commuters and retirement migrants, alongside communities that are in the 20 per cent most deprived within Scotland. Inequality and economic and social change are apparent, and the population is ageing with 25 per cent of residents over 65 in 2011.

Harris lies in the Outer Hebrides off the west coast of Scotland. Connection to the mainland is by plane (around two hours' travel), or ferry (around three hours' travel). Harris is sparsely populated, other than in the main town of Tarbert, and crofting townships are distributed around the perimeter of the island. Approximately 2,000 people reside in 911 homes, with 41 per cent being in one-person households. The population of Harris has steadily declined, falling by almost 50 per cent

Figure 1.4: Map of the study areas

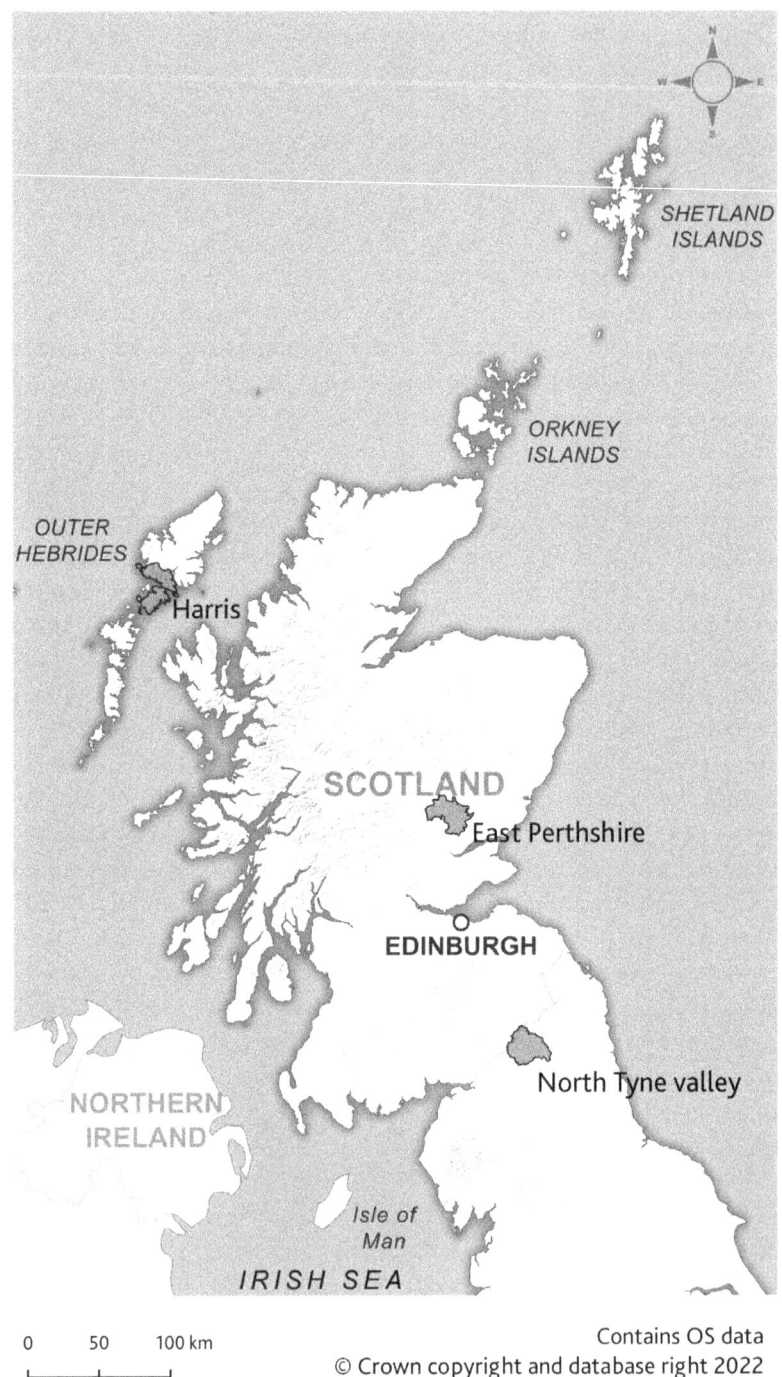

Contains OS data
© Crown copyright and database right 2022

since 1951. In 2018 32 per cent of residents of Harris were aged 65 and over. One of the key features of Harris today, in common with many of Scotland's rural areas, is the level of community land ownership: 70 per cent of people in the Western Isles live on community-owned land. The primary industries are tourism, with some fishing and crofting, and a heavy reliance on the public sector.

The North Tyne valley in Northumberland includes the four civil parishes of Bellingham, Kielder, Falstone and Tarset and Greystead in the north-west of the county, bordering Scotland. Combined, the four parishes cover an area of about 530 km² and had an estimated population of 2,019 people in 2017, one of the lowest population densities in England. Part of the area is within the boundaries of Northumberland National Park. The travel time by car to Newcastle is 80 minutes from Kielder, or around 50 minutes from Bellingham: it is largely viewed as beyond commuting distance from Newcastle. There is a wealth of history in the area, with many scheduled monuments, listed buildings and archaeological sites. Key industries include agriculture (hill farming) and forestry, with tourism and associated activities having become much more important in the last decade, particularly due to its International Dark Sky Park designation.

There were three stages to the fieldwork in each area. We firstly spoke with a range of 'gatekeepers', such as advice agencies, foodbanks, public bodies and third-sector organisations, who were likely to come across people experiencing financial hardship. We discussed similar topics with the gatekeepers in each of the areas, using pre-prepared topic guides based on our overall research approach and the issues arising out of our review of the literature. As a result of the flexible, semi-structured interview method, respondents could introduce unexpected topics and contribute important new insights to the conversations. All the interviews were recorded with the consent of the participants and subsequently transcribed for analysis.

The second stage of fieldwork consisted of interviews with individuals who had experienced financial hardship, or who were at risk of doing so. Sometimes these respondents were found with the help of gatekeepers, sometimes through 'snowballing' and on other occasions through placing requests in local newsletters or on community Facebook pages. These interviews explored people's life histories and experiences of living in their local area. We particularly sought to gather information on life events, work, housing, sources of support and other key issues pertinent to the study. Most of these interviews took place in people's own homes, although two respondents experiencing mental health challenges were interviewed in the premises of voluntary support organisations. Again, the semi-structured interview method empowered respondents to talk about their lives in ways which made sense to them, and which allowed them to introduce issues and insights which might otherwise have been overlooked. All these interviews

Table 1.1: Descriptive information about respondents in the three study areas

		Isle of Harris	East Perthshire	North Tyne valley
Gatekeepers	VCSEs	6	5	7
	Public	4	3	3
	Private	1	1	2
Gender of individuals (gatekeepers)	Male	2 (12)	2 (3)	2 (7)
	Female	3 (12)	3 (10)	2 (7)
Focus group themes		Tourism economy Community trusts Poverty/services Social stigma COVID-19's impact	Welfare support Health and care Housing Social isolation COVID-19's impact	Poverty/welfare Community trusts Health and wellbeing COVID-19's impact

similarly were recorded with the consent of the participants, on condition of confidentiality and anonymity, and subsequently transcribed for analysis.

Table 1.1 presents some descriptive information about the respondents in each of the three study areas. In each case there was a good gender balance among the individuals and among the gatekeepers interviewed, with ages ranging from people in their 20s to some in their 80s. There was some difference between the study areas in the employment status and housing tenure of the individual respondents: in Harris respondents were mainly retired, employed or self-employed; in Perthshire they were more likely to be unemployed, ill or in casual employment, and more often in social housing; while in Northumberland respondents included unemployed, self-employed and retired individuals.

The largest number of gatekeepers interviewed were voluntary and community organisations and social enterprises, ranging from those engaged in community development, housing and healthcare to those focused on providing advice, support and food to individuals and families in times of need. We also spoke to public sector organisations operating at a local level, and often to representatives from various departments. We spoke to few private businesses. Many of the gatekeepers kindly provided us with supplementary information after the interviews, and several helped us in finding individual respondents.

All the interviews with gatekeepers and individuals in Perthshire and in Harris were conducted face to face between September 2019 and February 2020, but face-to-face interviews were not possible for most of the interviews in Northumberland following the spread of COVID-19 to the UK and the announcement of the first national lockdown in March 2020. Instead, these interviews took place online (using video meeting software or occasionally by phone). Despite our worries these worked surprisingly well with high levels of rapport and trust evident and similarly full and frank conversations recorded. The only noticeable difference was the greater difficulty and delay in finding willing respondents.

The final stage of the fieldwork involved themed focus groups in each area, which were used both to test the validity of the findings from the earlier interviews and to explore how rural financial hardship and vulnerability had been experienced during the COVID-19 pandemic and first lockdown. Additional topics for the focus groups varied across the areas, based on the pertinent themes emerging from the interviews, as shown in Table 1.1. The focus groups in all three areas were conducted during the pandemic and so were carried out online. To make these effective in such circumstances, we followed the emerging advice for conducting fieldwork in a pandemic (Lupton, 2020) and restricted the number of participants in each focus group to a maximum of six. As a result, the focus groups were highly successful with rich material emerging. As before, all of these were audio recorded with the consent of participants and transcribed for analysis. The data were analysed thematically, drawing on the conceptual framework, following usual methods for qualitative analysis.

The validity of the findings was further tested, and endorsed, in spring 2021 through several presentations to national and local stakeholders. Quotes from the interviews and focus groups are included in subsequent chapters with codes where needed to identify the study area and whether this was an individual, gatekeeper or focus group, but no further details of the respondent are revealed to ensure anonymity in these small communities.

Structure of the book

This chapter has established that poverty and **financial vulnerability** affect people living in rural areas of Britain, as well as those in urban areas, but that this is much less recognised or addressed by policy or research. This book addresses this gap in our understanding of the processes underlying, and the experiences and impact of, low income and financial vulnerability in rural Britain in the period post-banking crisis, post-Brexit referendum and before and during the COVID-19 pandemic.

To this end, this chapter has also reviewed trends in poverty in Britain over recent decades and how these relate to changes in the policy context. The abandonment of Keynesian welfarism and its supplanting by neoliberalism, as well as the different priorities of Conservative and Labour governments, have been reflected in a sharp growth of poverty in Britain in the 1980s, a fall in poverty due to a combination of Labour's social policy reforms and economic growth after 1997 and further increases in poverty and precarity since 2010 under policies of austerity and welfare reform. Notable changes in the experience of poverty across these phases include a decline in poverty in old age, an increase in poverty in work and a transfer of social risk to individuals. These trends and policy changes have affected not only welfare benefits but also labour markets, housing costs, public services and the ability of local councils to ensure the wellbeing of their citizens.

In the next chapter we review existing knowledge from previous studies of rural poverty and social exclusion in Britain, Europe and North America. This enables us to develop a conceptual framework which focuses on the interconnections between individuals' and households' experiences and the structural and external processes bringing changes, for example, in local economies, employment, housing markets, welfare support and services, and how these are modified and intensified by place. It will also introduce many of the themes which recur throughout this volume, relating to work, welfare, governance, civil society, inequality, power and the potential challenges in extending the UK government's 'levelling up' agenda to rural citizens.

Chapters 3 to 5 draw on this conceptual framework to investigate these issues in three contrasting rural places – our study areas of East Perthshire, Harris and the North Tyne valley. Chapter 6 then examines in greater depth the impact of the COVID-19 pandemic and lockdowns on individuals in these three areas, and the responses to these impacts, while also drawing on other emerging studies of COVID-19's impacts in rural communities. These chapters form the empirical core of this book, presenting new and fresh evidence of the experiences of, and processes underlying, poverty and social exclusion in rural Britain.

The empirical findings from the three case studies, and the emerging themes of the book, are drawn together in Chapter 7. It begins by considering the four sources of support in our analytical framework (markets, state, voluntary and community sector, and family and friends) and how they act cumulatively to offset or reinforce social exclusion and financial vulnerability. This chapter also considers the role of place and its significance in understanding and addressing rural poverty.

Finally, the concluding chapter offers some reflections on the findings from this research and broader issues of social change in rural Britain. It also highlights some of the policy challenges emerging from the results and suggests some opportunities for policy development. More fundamentally, it reflects upon the power of the central state relative to the constraints facing local councils, raising issues of governance and democracy; upon rural–urban relations and the extent to which flows of wealth and capital shape rural poverty, prosperity and wellbeing; and upon the importance of voluntary and community effort, on the one hand in providing support for those in need and on the other in enabling and legitimising the abdication of the state's responsibility to its citizens in rural areas.

Poverty and social exclusion in rural Britain: a review

Research, media, policy and public attention have traditionally tended to focus on exploring the reasons for, extent of and responses to urban poverty, and particularly poverty experienced in the UK's biggest cities (Pacione, 2004). In contrast to other countries, such as the USA (Shucksmith and Schafft, 2012; Tickamyer et al, 2017), poverty in the UK has been widely viewed as primarily an urban phenomenon, associated with concentrations of poor housing, unemployment and social problems, with rural areas viewed in contrasting terms as idyllic and affluent (see, for example, Cloke et al, 2000a; 2000b; Milbourne, 2014). Indeed, Pacione (2004) argues that when the term 'rural disadvantage' entered policy debates in the UK in the 1970s, it suffered from a serious lack of credibility, with many arguing it was a contradiction in terms (McLaughlin, 1991).[1] Popular perceptions of rural areas and rural society at the time were generally characterised by images of an unchanging, comparatively affluent environment where the policy priority was the protection of the rural way of life. By contrast, deprivation was viewed as an urban phenomenon, characterised by highly visible poor housing, unemployment and dereliction (Woodward, 1996).

In addition to its urban focus, research on poverty has more often than not tended to take a fairly static and statistical data-focused approach, rather than being grounded in, or even incorporating, more qualitative studies of individuals' and households' lived experiences of low income (that is, limited monetary resources measured against a national level) alongside other factors such as poor health, low educational attainment, social isolation, precarious housing situations and cultural marginalisation (Lister, 2004; Milbourne, 2014; Milbourne and Coulson, 2020).

May et al (2020) noted that the focus on urban poverty remains (see also Williams and Doyle, 2016), with rural poverty tending to receive far less attention among UK academics and policy makers than urban poverty. Having said that, there has been an increasing recognition, in research terms at least, that some rural dwellers may suffer from similar challenges to those experienced in deprived urban neighbourhoods, including poor employment situations and a lack of appropriate and affordable housing, but that there are additional and specifically rural factors which may – often in combination – exacerbate the experience of poverty and exclusion in rural places. Research has also demonstrated that rural poverty and exclusion are more extensive and persistent than has often been assumed, although there is considerable variation in the extent of these challenges across different rural areas, and much depends on the indicators used to identify and measure the challenges, how they are used and at what scale.

Drawing mainly on UK-based literature, this chapter reviews and discusses the existing literature on rural poverty and social exclusion, beginning with a description of the extent of poverty and financial hardship in rural Britain and how different demographic groups are affected to different degrees. The chapter then reviews some of the specific factors contributing to rural poverty, before discussing the reasons why poverty and hardship often remain 'under the radar' in rural areas. In closing, we describe the analytical framework used in this study.

The extent of poverty and financial hardship in rural areas

Overall, rural populations have usually been shown to have somewhat lower poverty rates than their urban counterparts in the UK (Milbourne, 2016a), and it has traditionally been argued that rural poverty is confined to a small number of 'pockets of deprivation', or particular rural subgroups, rather than spread throughout the countryside (Bulman, 2017). A study of rural deprivation in case study areas of England in 1980 (McLaughlin, 1986; Bradley, 1987) and follow-up research in 1990 (Cloke and Milbourne, 1992; Cloke et al, 1994; 1995a; 1995b) found that around 25 per cent of rural households are living in or on the margins of poverty. In a more comprehensive study of the 7,000 households in the British Household Panel Survey (BHPS), published in 1998, one in three of those living in rural Britain were found to have experienced poverty between 1990 and 1996 (Chapman et al, 1998). However, spells of low income (defined as below half of mean household income) tended to be shorter than in urban areas, with the proportion of those who were 'persistently poor' significantly less than in urban areas. A recent report by the Scottish government (2021a) notes that income-based measures of poverty show that 15 per cent of people living in rural Scotland (170,000 people) are in relative poverty, compared with 20 per cent (850,000) of urban residents. The report also notes the lower levels of child poverty in rural areas compared with urban areas, with 19 per cent of children in rural areas living in relative poverty compared with 26 per cent of children living in urban areas. Analysis of Scottish data collected in the Poverty and Social Exclusion in the UK survey revealed significant poverty in all locations from urban to remote rural (Bailey et al, 2016). While, on most measures, poverty was found to be highest in the large urban areas and lowest in remote towns, remote rural areas tended to show higher poverty than remote towns. Rural incomes have also been found to be highly polarised, with the disparity between men's and women's earnings greater than the national average (Philip and Shucksmith, 2003; Scottish Government, 2018a; Atterton et al, 2019).

The main groups that have been found in previous studies to be experiencing poverty in rural areas are elderly people living alone (predominantly elderly widows) who generally have the state pension as their sole source of income, and low-paid manual workers' households. In Scotland, Shucksmith et al (1994, 2) found that 'a disproportionate number of the poor are elderly, and a disproportionate number of the elderly are poor', with persistent low income significantly higher for this group in both rural and non-rural areas (see also Commission for Rural Communities, 2006a; Children in Wales, 2008). The Shucksmith et al study also found that, as in England, the other main element of rural poverty derives from the disproportionate number of people in low-paid occupations, notably in agriculture and tourism. Often, the self-employed were also found to have low incomes, as were single-person households. In some remote areas in Scotland, such as Harris, almost the whole population was considered as a low-income group. In general, therefore, rather than concluding that pockets of disadvantage exist, as had been the traditional assumption regarding rural poverty, poverty was revealed to be rather widespread in rural Scotland (Shucksmith and Chapman, 1998). The analysis also confirmed the role of migration in affecting poverty levels in rural areas, in that the rich were generally moving into rural areas, while those on low incomes moved out, often not through choice.

While the incidence of persistent unemployment has been found to be less in rural areas, the incidence of persistent low pay can be significantly greater (Chapman et al, 1998; see also Children in Wales, 2008; Milbourne, 2011; Ray et al, 2014; May et al, 2020). Research in Wales found that 29 per cent of low-income households contained at least one person in work (Children in Wales, 2008). This research also revealed the high reliance of poor households on the private sector for the provision of key goods and services, and on private transport due to inadequate public transport provision. This means that individuals and families often make major sacrifices in other aspects of their lives so they can own and run a car.

The relatively low escape rate from low pay for individuals employed in small rural workplaces, combined with the dominance of microbusinesses in rural areas, suggests that the lack of opportunity to change jobs to work for larger employers may be an important explanatory factor (Chapman et al, 1998). This finding was echoed two decades later by Shucksmith (2018a), who suggested that poverty in work because of low pay is typical of rural areas in many countries. He argued that this is compounded by poor access to transport, services, employment and training, and also by a reduced tendency for rural people to seek welfare support to supplement low pay. This is commonly to avoid any stigmatising perception of 'welfare dependency', often combined with poor access to information and advice

(Shucksmith, 2018a; see also Williams and Doyle, 2016). As noted by Satsangi and Wilson (2020), these challenges are a particular feature of more remote rural locations from which residents are not able to commute to well-paid jobs in urban centres, as is the case for at least some people who reside in more accessible rural locations.

Certain groups in the population are more likely to experience multiple deprivation, regardless of whether they live in a rural or an urban location (Pacione, 2004). These groups include the unemployed, workers on low pay, part-time workers, people with limited disposable income such as the young, pensioners, elderly men and women, lone parents and people suffering ill health or disability. Pacione's work also acknowledged the importance of contextual factors in conditioning the nature and impact of disadvantage, with the poor in rural areas disadvantaged by limited accessibility to services, due to the friction of distance and inadequate public transportation.

Echoing the findings of Shucksmith and Chapman (1998; see also Shucksmith et al, 1994), analysis of BHPS data from 1991 to 2008 found that rural poverty is not a rare experience affecting only a minority of people. On the contrary, 50.2 per cent of rural households experienced poverty at some time during this 18-year period, compared with 55.2 per cent in urban Britain (Vera-Toscano et al, 2020). This analysis revealed a disproportionately high risk of persistent poverty among women and those aged 65+. That half of the rural population experienced poverty at some point during this longer period of 18 years is consistent with the earlier finding by these authors that around a third experienced poverty during the shorter period of 1991–96 (see Chapman et al, 1998). Palmer's (2009) analysis of official data on low income in England indicated that a lower proportion (19 per cent) of households in rural England were living in poverty in 2006–07 (cited in Milbourne, 2014, 568). In rural Wales Milbourne's (2014) study found a similar proportion (18 per cent) of households living below the official UK poverty threshold of 60 per cent of the national median income in 2007. However, studies in Scotland have confirmed high levels of financial stress in rural areas, with fewer than half of those in the Highlands and the Western Isles 'coping well or very well financially' (Scottish Household Survey, 2012), and particular challenges with widespread fuel poverty in rural areas (Scottish Government, 2018b).

While much of the previous research has revealed the particular prevalence of poverty among the older-age population, it is also important to discuss the experiences of poverty among rural young people, and the ways in which these experiences hamper their long-term life chances. In 2000 the Joseph Rowntree Foundation's Action in Rural Areas Programme found evidence of young people being integrated into two very different labour markets: the national (distant, well paid, with career opportunities) and the local (poorly paid, insecure, unrewarding and with fewer prospects)

(Joseph Rowntree Foundation, 2000). Level of education and social class were the two key elements which enabled some young people to access the national labour market, while others did not: those lacking educational qualifications effectively became 'trapped' in the local labour market with fewer opportunities for employment, education or training leading to reduced life chances (see also Bailey et al, 2004). This research also found a particularly strong interplay between employment, housing and transport. Often, young people could not afford to live independently and run a car on the generally low rural wages, but for many young people a car was a prerequisite to having a job, and particularly a better-paid job outwith their local area.

Work by Shucksmith (2004) and Jentsch and Shucksmith (2003; 2004) explored the ways in which the life experiences of young people have been changing because of processes such as globalisation, the rise of the 'risk society', increasing individualisation and ever more complex and protracted youth transitions. Their research found that young people in rural areas often found themselves and their needs particularly invisible, that they are denied spaces for social interaction, that they have particularly poor access to transport – with the cost of private transport a particularly significant barrier to finding and maintaining employment. This exclusion of young people from full participation in rural society, particularly through their lack of access to public spaces, resonates with the concept of rural citizenship. These kinds of exclusionary practices actively reproduce hegemonic and conservative visions of rurality and certain expectations of citizenship, thereby effectively excluding some groups from full participation (Yarwood, 2017). As Painter and Philo (1995) argue, if people cannot be present in public spaces without feeling 'out of place', it is hard for them to consider themselves full citizens. Young people living in rural areas also often face additional challenges, including: inadequate careers advice; lack of access to affordable housing (particularly in rural areas also experiencing high levels of second and holiday home ownership, and which are 'retirement hotspots'); a reliance on poor-quality housing which is less energy efficient and more expensive to heat leading to higher instances of fuel poverty (see also McKee et al, 2017); and the disappearance of support services, including support provided by 'traditional' institutions such as family and the church.

Although there are benefits to living in rural areas for (at least some) families, including opportunities for children to play in the open air independently, relatively low crime levels and local leisure activities (for example, Young Farmers' Clubs), children living in households in Wales without a car (which are more prevalent among those on low incomes) experienced a lack of access to services (including for health, leisure and so on) and negative impacts on their social networks as public transport services were often poor or non-existent, unreliable and/or expensive (Children in Wales, 2008). In some rural areas, the Welsh study also found a lack of provision of, and access to, affordable childcare services (see also Little and Morris, 2002; Rural Services Network, 2014; Public Policy Institute for Wales, 2016), a high level of housing need and homelessness (including for families with children) and negative impacts on the lives of children whose parents are faced with the challenges of low pay and unstable employment. These challenges were especially acute for migrant worker and minority ethnic families and for families with children with disabilities, on low incomes and/or in which domestic abuse was occurring.

More recent research in a sparsely populated rural area of northern Britain found that the poverty challenges faced by young people before the 2008 financial crisis persist but are also being exacerbated by several new factors (Black et al, 2019). These relate to poor digital connectivity and transport services, fewer opportunities to secure a reasonable livelihood locally, the stigma attached to claiming welfare support (which seems to have become a more significant barrier over time), the changing nature of the labour market with a shift towards less secure forms of employment and reduced provision of services and welfare reforms. With reference to youth studies, Bourdieu's theory of practice, concepts of welfare regimes and welfare mix and the impacts of the crisis and associated austerity policies on the distribution of social and societal risk, Black et al (2019) argue that the overwhelming (and increasing) reliance of young people on family for support (to fill growing gaps in social protection provision) generates further inequalities through what is termed 'secondary impact austerity'. Young people feel directly and unevenly the economic effects and policy changes which impact on their parents' and wider communities' ability to offer them support. Thus, the new factors described previously exacerbate the transfer of social risk and the deepening of poverty for vulnerable groups. This was worsened in Black et al's rural study location by the shared moral imperatives (local habitus) which stigmatise access to state and charitable support. Because young people in rural communities rely heavily on personal networks for securing employment and accessing other forms of support, it is easy for those from families on a low income to become stigmatised due to their higher visibility in the community (Glass et al, 2020).

Returning to experiences of poverty and deprivation among the rural population generally and bringing this review of the extent of rural exclusion and financial hardship up to the present day, as May et al (2020) argue, there has been relatively little written about the impacts of austerity on rural areas, despite a number of calls for more work that contextualises studies of poverty in the local social and cultural contexts that shape the everyday experience of those living it (see, for example, Milbourne, 2004; 2014; Cloke et al, 2007). Milbourne and Coulson (2020) also note the impact of austerity on poverty, first in relation to cuts to public sector services and funding, which in turn have led to cuts for the voluntary sector which has at the same time gained a more important role in service provision (see also Shucksmith, 2016). Work by Wilson and Copus (2018) has also revealed the acceleration of public service cuts in rural areas in Scotland and the extent of population decline in some more remote areas. The second element of austerity relates to major reform of the benefits system, not least in relation to the roll-out of Universal Credit (UC). This reform has tended to lead to a societal shift in terms of how poverty is framed as an 'individualistic' problem related to the 'dysfunctionality' of households and individuals, which may reinforce the moral discourses of rurality that prioritise self-reliance over state-provided welfare support (see also Milbourne, 2016b). Milbourne and Coulson (2020) also highlight the evidence showing how austerity policy and welfare sanctions have led to a dramatic rise in the number of households experiencing food poverty across the UK and requiring emergency food aid (see also Loopstra et al, 2015; Beck et al, 2016; Lambie-Mumford and Green, 2017; Scott et al, 2018).

May et al (2020) add to our understanding of the impacts of austerity through their own study of the geography of austerity and food banking in rural England and Wales, and how these new geographies are overwriting and compounding problems of rural poverty. They found that cuts to local authority spending and to welfare benefits have hit urban areas hardest but have hit the most deprived rural local authorities disproportionately and worsened deprivation in those areas by particularly affecting low-income households. They also found that some changes which have a direct and rapid impact on food security (such as benefit sanctions) are more marked in rural than in urban areas (at the same time that welfare assistance schemes have been cut back, particularly in rural areas). Their work explores the scale and uneven distribution of this aid across rural areas and discusses some of the challenges common to those experiencing both food shortages and poverty in rural areas, including: transport and service deprivation, which has been exacerbated by cuts to bus routes, the consolidation and closure of job centres and Citizens Advice centres – which are also the organisations that make referrals for other services, for example, foodbanks – other services such as youth, children and family centres and libraries run by the local authority; the 'digital by default' design of UC and other benefit claims; the rural premium (that is, higher spending on everyday goods and services, which is particularly challenging for those on low incomes); stigma; and shame. While the latter is a reason for people to avoid using foodbanks in any location, it seems to be exacerbated in rural locations, where there is a sense in which not being able to feed your family contravenes the usual expectations of the self-reliance of rural communities.

Particularly interesting in the context of the study that forms the focus of this book, May et al (2020) explore how austerity and narratives of dependency and deservingness work out differently in two very different rural places (working-class communities in the South Wales Valleys and conservative coastal and farming communities in South-West England) to demonstrate how local cultures of poverty, welfare and charity (Cloke et al, 2007) continue to shape the variegated geographies of austerity and responses to it. May et al's work also explores the varied geographies of rural food banking and notes the challenges that those who need to access a foodbank in rural locations experience (similar to the challenges applicable to those seeking to access fresh and affordable food in supermarkets), including distance, and a lack of affordable and reliable public transport to access a referral centre or the foodbank itself. This work also notes the strategies and solutions that have been put in place by providers to overcome some of these challenges for service users, including giving out larger parcels to reduce the frequency of travel required, working with other service providers to make parcels available in other locations and even delivery services (though these can result in additional demands on and expenses for volunteers).

Concerns about a 'cost of living' crisis emerged in late 2021 and have gathered pace in 2022 with consumers experiencing rapidly increasing prices in particular for food and fuel as a result of various factors, including the COVID-19 pandemic and Russia's invasion of Ukraine, both of which have led to supply chain challenges and thus food and energy shortages and rising prices. Research by Robinson and Mattioli (2020) found that it is rural households which exhibit the greatest 'dual energy vulnerability' and are therefore likely to suffer the greatest financial pressure as the cost of living crisis deepens. It is also the case that those rural households that are facing ever higher energy prices are consequently unlikely to be able to afford to undertake additional measures to improve the longer-term energy efficiency of their homes with the related financial savings that would be achieved in the longer term. Action with Communities in Rural England (ACRE) commented in May 2022 that the UK government is not doing enough in terms of providing substantive support to stave off the cost of living crisis which is leading many more rural households to have to choose between putting fuel in their car, heating their house or providing a hot meal for all of the family, and which is threatening the sustainability of rural communities (ACRE, 2022).

This section has summarised the key findings of studies of rural poverty over the last 20–30 years. This work has helped to demonstrate the significant challenges of poverty, deprivation and exclusion in many rural communities and dispel the myth that poverty is only an urban phenomenon. The review now moves on to discuss in more detail the factors contributing to poverty and financial hardship in rural areas, and the reasons why rural poverty still remains relatively hidden and under-reported.

What are the contributory factors for poverty and financial hardship in rural areas?

Some of the reasons why households and individuals experience financial hardship and poverty apply irrespective of geographical location. These factors might include economic change, recession and slow economic recovery, a reduction in job security, declining welfare and other services and associated austerity measures, changing family and social relations and a decline in established institutions alongside the rise in individualistic values and attitudes across society which have led to a redistribution of risk to citizens and away from the state as its responsibilities shrink and its relationships with citizens are altered (see, for example, Shucksmith and Chapman, 1998; Philip and Shucksmith, 2003). Such factors have often been found to impact most severely on already disadvantaged individuals and households, including older people, lone parents and people experiencing in-work poverty (Asenova et al, 2015). However, as argued by the Public Policy Institute for Wales (2016), while there are some similarities between rural and urban poverty, particularly in relation to the effect of poverty on the individual, their scale and causes can often differ. Moreover, there are some types of deprivation which are more prevalent in rural communities than urban communities. Several other factors have been found to be particularly important in contributing to financial hardship and poverty in rural locations, which can be complex and multidimensional (Scott et al, 2007; Bertolini et al, 2008). This section considers these factors in turn.

Curtin et al (1996) suggest that the historical dominance of agricultural production in rural areas has created differences in the generation of poverty compared with urban areas. For older people who were employed on low wages in the agriculture and related sectors, one of the key 'risk factors' explaining their situation is that they were in jobs with low incomes during their working life, commonly with no occupational pension scheme. Such low-wage jobs, with limited opportunities for progression, are particularly prevalent in rural areas, often in primary sector activities.

Rural areas also have a distinctive organisation of space, namely sparsity of population and spatial peripherality, which may also generate distinctive manifestations of poverty (see Philip and Shucksmith, 2003). Shucksmith and Chapman (1998) describe the structural forces of change in many rural areas which have led to considerable economic and social change, including an ageing population and changing and socially selective migration flows. Fundamental demographic change is occurring in many rural areas as populations age more rapidly than those in urban areas, due to people ageing in situ, older people moving in (at either retirement or pre-retirement life stages) and continued youth outmigration. While the majority of older people moving into rural areas are relatively wealthy, this has contributed to artificially raising house prices in many rural places, leaving housing beyond the reach of many local residents, and particularly younger people who are reliant on low-paid, local employment. This is combined in many rural areas with a shortage of adequate housing to rent, either privately or in the social rented sector, often due to many rental properties being taken out of the local housing market and becoming accommodation for visitors on a short-term basis (see Willet, 2021 describing experiences in Cornwall), leaving some individuals and households 'sofa surfing' and staying with friends and family in inadequate circumstances, often for long periods of time.

Important economic changes have also led to significant social change, including globalisation and the shift away from primary production (including agriculture and forestry) to service sector employment (see also Shucksmith, 2000). As Pacione (2004, 387) argues, the effects of economic globalisation and the marginalisation of some less-favoured areas may intensify the problems of deprivation experienced for remoter rural areas and populations. For example, in terms of the evidence for financial hardship among farmers, the Food Research Council in 2014 estimated that 25 per cent of all UK farmers were living in poverty, and this is likely to be due to the changes in the farming sector, including reductions in labour (Food Research Council, 2014; see also Shucksmith and Chapman, 1998). The Food Research Council study also noted the challenge with low wages in the food production industries with workers often unable to afford to eat the food that they pick or pack (Food Research Council, 2014).

Rural areas are sometimes regarded (positively) as having relatively flexible labour markets with opportunities for short-term work and self-employment for those who wish to take advantage of them, resulting in generally lower levels of unemployment and higher rates of employment and economic activity (see Scottish Government, 2021a). However, for many people, this flexibility can contribute to their experiences of financial hardship and poverty. These labour market features include the persistence of low pay and unstable and seasonal and/or self-employment, a predominance of part-time working and underemployment with people reliant on low-pay and low-skill jobs with limited career progression opportunities, and fragile economies with fewer opportunities for employment as well as training and/or career progression (see Commission for Rural Communities, 2006b; Public Policy Institute for Wales, 2016; Scottish Government, 2021a). Added to this is often poor public transport and the prohibitive expense of private transport which leaves people reliant on fewer local opportunities, and the divisive effects of 'word of mouth' and informal methods of recruitment and job search (as opposed to formal job search strategies), which leaves those excluded from such networks unable to access opportunities (Lindsay et al, 2003). A lack of (affordable) childcare and/or eldercare in rural places may also act as a barrier to employment (see Shucksmith, 2000; 2016; 2018;

Bailey et al, 2004; Public Policy Institute for Wales, 2016). There is also a challenge associated with tied housing – housing provided with a job – which is often found in rural areas. This means that people are unable to change jobs due to low pay or poor conditions, as this would mean losing their house too.

Fuel poverty is another characteristic of rural areas (especially remote rural and island communities) due to the higher price of fuel for off-grid properties that are often reliant on electric storage or oil-powered heating. Due to the generally older and larger rural housing stock, more fuel is required on average to heat properties (see, for example, Skerratt and Woolvin, 2014). The Scottish government (2021a) notes that a third of households in remote rural areas were in extreme fuel poverty in 2019 compared with 11 per cent of households in the rest of Scotland. Simcock et al (2020) found particular groups in the population are at risk of experiencing fuel poverty, including low-income households (including older people, lone parents and those with health conditions) and those living in properties that are older and less energy efficient. In addition, those living in the private rented sector may experience fuel poverty as they have less opportunity to take up energy-efficient appliances and building improvements because they do not own buildings and fittings. Simcock et al (2020) also note that recent research has uncovered new groups experiencing fuel poverty, including young people, and they particularly note the vulnerability of rural households to fuel poverty due to the greater dependence on more expensive non-gas heating fuels. The authors' work also highlights that households with low incomes are at risk of transport poverty, as are households with members who have mobility problems and rural households who lack access to alternatives to car use. Those households that are vulnerable to both forms of poverty include low-income households, households with children and those from ethnic minority communities. For those rural households that lack reliable and quick digital access, adequate digital skills and/or the income to afford the necessary equipment, experiences of poverty and exclusion may be particularly acute (see Public Policy Institute for Wales, 2016).

Other evidence (see, for example, Citizens Advice Service, 2018) suggests challenges of accessing affordable food with reasonable shelf life and the unavailability of certain foodstuffs (including quality fresh produce) in remote rural areas. Some remote rural respondents in this study expressed the wish for more locally sourced foodstuffs to be available. These challenges are supported by Corfe (2018) who found that 26 per cent of rural areas in the UK could be classified as 'food deserts' compared with 17 per cent of urban areas, with individuals with restricted mobility (for example, due to age, physical ability or cost) facing additional difficulties accessing affordable food stores.

Declining service provision is a significant contributory factor for social exclusion in rural areas (see, for example, Public Policy Institute for Wales,

2016), including diminishing availability of local welfare advice and other services (as provision is centralised or moved online) and reduced public transport services. The decline of such services leaves people either without access to any services or with long distances to travel to reach the nearest market town where services are available, either by public transport (if it is available) or by more expensive private transport (see also Children in Wales, 2008). For example, a study of poverty and social exclusion in the Argyll and Bute local authority in Scotland in 2004 revealed high levels of access poverty, as would be expected in a rural local authority, but considerably lower levels of education deprivation (even so, the area contained substantial numbers of people in deprivation according to the Scottish Index of Multiple Deprivation [SIMD] measures) (Bailey et al, 2004). The closure of libraries and other meeting spaces in rural areas over recent years because of public sector funding cuts reduces the opportunities for local people to come together, potentially contributing to social exclusion.

These challenges are all experienced in a geographical context where households often face higher living costs, and these tend not to be taken into account in income and deprivation measures. The Public Policy Institute for Wales (2016) refers to this as the 'rural poverty premium', where the poorest in society often have to pay more for essential goods and services than those who are better off; this might be because they lack digital access to 'shop around' for the best deals; because they are geographically distant from larger centres where they would be better able to take advantage of more competitive prices for food, fuel and so on (Hirsch, 2013); because the closure of bank branches and (free) bank machines means that they are unable to take advantage of mainstream financial services or to take out cash regularly; and because they need to pay more to heat their (usually larger and less energy-efficient) homes.

The Minimum Income Standards work – which has been carried out across the UK (see, for example, Smith et al, 2010) and in specific areas, such as the Highlands and Islands (see Hirsch et al, 2013) – has revealed that, with some limited exceptions (such as leisure activities for primary school children), overall, rural households face additional costs. The additional costs for a family with two children were particularly large in rural areas when compared with urban areas, though they were also larger for pensioner and single working-age person households too. Transport was found to make up the most significant proportion of these additional costs (Smith et al, 2010) – between 60 per cent and 100 per cent of the difference – reflecting the need for rural households to have a private car. Domestic fuel costs were found to make up the next largest proportion of the difference, with other items such as food, household goods and social participation making up smaller proportions of the difference (Smith et al, 2010).

Hirsch et al (2013) found that households in remote rural Scotland require significantly higher incomes to attain the same standard of living as those living in other parts of the UK (between 10 per cent and 40 per cent higher, with additional costs potentially exceeding 40 per cent for households in more remote island locations). These premiums are greatest for single people and families supporting children, and can be even higher (above 40 per cent) for those living in the most remote island locations. There are three main reasons for this premium: higher prices for food, clothes and household goods, higher household fuel bills and the longer distances that people have to travel routinely, particularly for work.

In the 2016 update to the earlier work, a similar pattern was observed (Davis et al, 2016). This update found that the minimum acceptable standard of living in remote rural Scotland typically requires between a tenth and a third more household spending than in urban parts of the UK (the lower price of petrol and diesel in 2016 had reduced the additional cost for people having to travel long distances, particularly regular travel for work). Davis et al (2016) argue that a framework for addressing these higher costs needs to consider issues around energy costs, shopping costs and travel costs in a joined-up way, which takes account of the influence of local infrastructure and the development of communities and jobs (Hirsch et al, 2016). The Scottish government (2021b) has identified additional minimum living costs for households in remote rural Scotland that typically add 15 to 30 per cent to a household budget, compared with urban areas of the UK (Table 2.1). More broadly, the 2021 update to the Minimum Income Standard (Davis et al, 2021) argues that millions of people in the UK risk falling well short of the Standard because they lack secure work or decent pay and benefits, particularly due to the removal of the £20 a week uplift in UC in October 2021. As the UK emerges from the COVID-19 pandemic, but is faced with a growing cost of living crisis, this underlines the importance of both stable opportunities to earn a decent income and improved support from the state to help people who are being held back from an acceptable living standard. Davis et al (2021) argue that, as the UK rebuilds its economy in the wake of the pandemic, the government should prioritise the creation of decent jobs that provide stability, pay at least the real living wage and give people options about working hours.

Many of the factors described here are mutually reinforcing. For example, as service provision declines in local rural areas, those who cannot afford to run their own car – including young and older people – may find that they can no longer access even basic services, including advice and support. The stigma that is often associated with claiming benefits may leave rural dwellers more heavily reliant on friends and family and social networks for support; as these social networks decline, these individuals and families are at even greater risk of falling into deeper, more persistent and even more hidden poverty

Table 2.1: Percentage of additional Minimum Income Standard costs in remote rural Scotland in 2021

Household type	Remote rural Scotland	
	Mainland	Island
Couple+2	16.2%	15.5%
Family with children, rounded uplift (based on couple+2 case)	**16%**	**15%**
Single working age	20%	13.6%
Couple working age	20.6%	13.8%
Working age rounded uplift (based on average of single and couple)	**20%**	**14%**
Single pensioner	30.5%	37.1%
Couple pensioner	20.7%	29.1%
Pensioner rounded uplift (based on average of single and couple)	**26%**	**33%**

Source: Scottish Government (2021b). Contains public sector information licensed under the Open Government Licence v3.0.

(see Lichter and Graefe, 2011; Tickamyer and Henderson, 2011; Sherman, 2013). At the same time, as Bailey et al (2016) argue, people in poverty, no matter where they live, report feeling lower levels of social support and feel less able to turn to family or friends to help with practical or personal problems. They also take part in fewer social activities and have less frequent contact with friends. For poor adults, Bailey et al (2016, 6) found that people living in remote rural areas appear to have particular problems with low levels of support. They argue that this fits with previous literature which has found that poverty in rural areas may be more isolating in its impact, due to the greater visibility of individuals within rural communities and a rural ideal of self-reliance. Added to this is the lack of recognition of these challenges from urban dwellers – and often from urban-based policy makers and media – who regard the countryside as a 'rural idyll' (Lowe et al, 2012).

Researchers have sought to organise these factors into groups. For example, a report by the Commission for Rural Communities in 2006 designated four groups: financial poverty (referring to no wages, low wages or small pensions); access poverty (referring to the challenges of accessing transport and services); network poverty (referring to the lack of informal contact with, and help from, friends and neighbours); and attitudes and perceptions (referring to the belief in the rural idyll which prevents recognition of others' disadvantage).

In summary, while there are wider societal processes and changes which contribute to experiences of financial hardship and poverty, no matter where an individual or household is located, there are many factors which are particularly important in a rural context and which lead to the substantial proportions of rural people experiencing these situations. Powell et al (2018) identify five dimensions of rural poverty associated with housing needs, access to services, public transport, fuel poverty and rural economic development. Many of these factors also lead to rural poverty and exclusion remaining relatively hidden and 'under the radar'; this is explored in the next section of the chapter.

Why are poverty and financial hardship often hidden and under-reported in rural areas?

Researchers have long argued that poverty and financial hardship are relatively 'hidden' in rural areas and less visible than they are in urban centres (see, for example, Shucksmith, 2000; Pacione, 2004; May et al, 2020) and that this may be due to a variety of different reasons.

It is argued that the existence of the 'enduring myth' of the power-infused and backward-looking rural idyll (Cloke et al, 2000b; Pacione, 2004; Shucksmith, 2018b) creates a particular social construction of the countryside which strongly influences perceptions of rural life. In a study carried out by Shucksmith et al (1996), the majority of respondents from rural areas presented rural society as inherently good, caring, safe and advantaged, while urban society was perceived as inherently degenerate, dangerous and disadvantaged. One key characteristic of this rural idyll is the more limited infrastructure, which is part of the appeal of living in a rural place but also a key driver of rural poverty, as described in the previous section (Cloke et al, 1995b).

This links closely with the concept of 'rural citizenship', a status that individuals are denied if they do not have access to the infrastructure that they need to participate fully in society, including welfare and employment services. Effectively, the rural idyll actively excludes some individuals and groups, including those experiencing poverty but also women, ethnic minority groups and those with disabilities, from being full and active citizens (Painter and Philo, 1995; Halfacree, 2007; Yarwood, 2017). In recent years the withdrawal of the state from providing many services in rural locations, and the subsequent absence of service provision and/or heavier reliance on the voluntary and community sector, have further eroded the citizenship of rural dwellers who are in poverty or in receipt of welfare payments and who are no longer able to access these services in the same way as urban citizens (Tonts and Larsen, 2002). Building on this, Cresswell (2009) argues that citizenship relies on 'prosthetic' materials, such as shops, services, employment and transport, to achieve full societal and welfare rights. The closure of a post office or local welfare office leaves some rural citizens lacking the support they need to fully participate as citizens of their wider society.

A further element of the rural idyll relates to how rural people subjectively view their situation in contrast to how it may be objectively measured, combined with a stronger sense of the self-sufficiency of traditional rural households and mutual support from idealised rural communities. For example, in Shucksmith et al's work (1994; 1996; Shucksmith, 2000), rural respondents looked back to a point in their past where poverty had been much more commonplace and obvious; in effect, rural people compared their current lifestyles to those of the past which were much harsher and to those of their urban counterparts, rather than to the lifestyles of the majority of people around them.

Moreover, personal characteristics and differing individual adaptation levels mean that contrasting evaluations may be made of the same objective environment (Pacione, 2004, 379). For some people, it is just accepted that certain difficulties, such as a lack of employment choice, are inevitable aspects of rural life (Cloke et al, 1994) and a 'price worth paying' for the benefits of rural living (Cloke and Little, 2005; Shucksmith, 2016). In Pacione's (2004) study, rural people reported being 'rich in spirit, but poor in means' (see also Shucksmith et al, 1994; Philip and Shucksmith, 2003), with many residents placing a high value on the non-monetary aspects of rural life, such as low levels of crime, higher levels of wellbeing and quality of life, strong traditions of community solidarity and self-help (Sherman, 2006) and high-quality landscapes and scenery (see also Cloke et al, 1994). As a result, they may regard themselves as having a reasonably high quality of life and may not feel entitled to claim support (Cloke et al, 1994; Shucksmith et al, 1994; Shucksmith, 2000; Milbourne, 2014).

Research on poverty in rural Wales also found that people on low incomes tended to construct their lives more in relation to their social and cultural worlds than to issues of low income and material deprivation (Milbourne, 2014). This work revealed important disconnections between material and socio-cultural aspects of rural poverty, with community belonging and attachment to landscape appearing more significant than material hardship and social exclusion within the narratives of everyday life of those who were 'poor'. Milbourne (2014) also concludes that community belonging is bound up with particular moral discourses of welfare and rurality that act to perpetuate situations of material poverty within rural places.

Thus, the notion of the 'rural idyll' effectively tends to both obscure and exacerbate problems of rural poverty, as well as add to the stigma felt by those who experience it (see, for example, Cloke and Milbourne, 1992; Shucksmith, 2000; 2018; Milbourne, 2004; 2014; 2016a; Woods, 2006). The idyll also plays a key role in attracting wealthy incomers who then seek to preserve and even enhance it by constructing rurality in particular ways. In practice, this may mean opposing plans for new (affordable) housing locally, for example (see also Halfacree, 1993; Shucksmith, 2012). This process of wealthy incomers displacing less affluent locals has been termed 'spatial apartheid' or 'rural gentrification' by different authors (see, for example, Joseph Rowntree Foundation, 2000). Willet (2021, 121–2) also discusses the importance of the rural idyll in relation to the attractiveness of Cornwall in South-West England to tourists: 'Poverty ... is not a part of the expected Cornish tourist experience, and so consequently tourists are not directed towards the areas which challenge the rural idyll'. And she goes on: 'rural poverty and deprivation mean that the lived reality of many is very different from the visitor experience and notes that when viewed through the lenses of visitor assemblages, practices of deep poverty become reinterpreted as "romantic" – failing or unable to see the pain in them'.

Added to the challenges associated with the rural idyll, there is a set of related data issues. One issue is that the indicators that are often used to highlight poverty and therefore target resources are often not appropriate in a rural context. For example, the number of people living in flats or in social rented accommodation or the number of people officially classified as homeless (or indeed the number of homeless shelters or soup kitchens, that May et al [2020] term the 'classic institutions of poverty') are very much features of urban deprivation rather than rural deprivation (Gloyer, 2002). Conversely, car ownership, which is usually higher in rural areas due to inadequate (or a complete lack of) public transport, is not an appropriate indicator of relative prosperity. It is also sometimes the case that low-income poverty measures do not take account of cost of living differences which can be significant, particularly in remote rural areas (as described earlier; see also Scottish Government, 2021a).

Importantly, as May et al (2020, 410) argue, since rural poverty is, by definition, more dispersed than urban poverty (which is traditionally associated with housing estates), a focus on patterns of concentrated poverty tends to radically underestimate the extent of poverty in rural areas, if not render it completely invisible when there are low population numbers and dispersed settlement patterns with those experiencing poverty spread among relatively affluent households (Milbourne, 2014; see also Skerratt and Woolvin, 2014; Public Policy Institute for Wales, 2016; Joseph Rowntree Foundation, 2000). Place-based measures of poverty, such as the Index of Multiple Deprivation (IMD) in England and the SIMD, are generally accepted to underestimate the scale and extent of rural disadvantage and to be better suited to identifying and measuring urban deprivation (McKendrick et al, 2011). Even where such indices include a domain relating to access and distance, this is outweighed by the other domains which favour urban aspects of deprivation (Shucksmith, 2016). In addition, such indices usually incorporate benefit claimant rates as a proxy indicator, and previous research has found claimant rates to be lower in rural areas (this is discussed in more detail later in this chapter). In short, those experiencing poverty in rural areas simply do not have the visibility accorded to their counterparts in the city (Cloke et al, 1997; Shucksmith, 2000; Milbourne, 2004; 2014; May et al, 2020).

The finding of Bailey et al's (2004, ii) study of deprivation in Argyll and Bute perhaps summarises the geographical distribution of poverty in rural areas particularly well:

> A focus on deprived areas alone can give a misleading impression of the distribution of deprivation across the area. Deprived areas are concentrated into the larger urban areas but deprived individuals are found in urban and rural areas across the authority. There are some locations, however, which do have significantly higher levels of need than average. Dunoon, Rothesay, Campbeltown and Islay, in particular, have both high concentrations of deprived individuals and large numbers in absolute terms. Some smaller settlements and islands also have high concentrations though absolute numbers involved are much smaller.

While in urban centres those in poverty may rely heavily on public sector-provided services and thus be counted in official poverty measures, many in rural areas may not have access to such services and are therefore not counted, while also being forced to rely on informal support from friends and family or private sector services, which may be more expensive (Cloke et al, 1997; Milbourne, 2004). There is also a challenge with accessing some data in rural areas, particularly relating to income and employment, at small enough scales due to anonymity and confidentiality concerns.

Finally, evidence suggests that rural dwellers may be less likely to claim benefits if they are experiencing poverty and hardship than their urban counterparts (see, for example, Shucksmith et al, 1994; 1996; Bramley et al, 2000; Commission for Rural Communities, 2007a; 2007b; Pugh et al, 2007). Research for Commission for Rural Communities (2007a) analysed the UK Department of Work and Pensions' (DWP) Family Resources Survey data linked with actual administrative data on Pension Credit uptake and found a statistically significant difference: overall take-up was lower in rural areas with 42 per cent of those eligible in rural areas failing to claim compared with 35 per cent in urban areas. Non-claimant rates were much higher (54 per cent) in villages and isolated dwellings. As benefit claimant levels may be used as an indicator of economic and social need, rural needs may therefore be underestimated. There may be several reasons for the lower claimant levels in rural areas.

Most research has suggested that many people in rural communities feel a stronger social stigma attached to claiming benefits (perhaps due to values which tend to be associated with rural areas such as a strong work ethic and self-reliance), whether they are required to do so at a potentially intimidating benefit office (which may be costly to travel to in a town centre by public transport or private car) or at a local village post office (see, for example, Rank et al, 1993; Shucksmith et al, 1994; 1996; Shucksmith, 2001; Sherman, 2006; 2009; 2013; 2021; though this conclusion is not supported by all research: see, for example, Bailey et al, 2016, iii). In short, it is usually argued that there is more visibility and less anonymity – and therefore more potential for stigmatisation – when claiming benefits in a rural location (Rank et al, 1993; Shucksmith, 2001). This is one of the more negative implications of having a strong sense of community and, related to this, a sense in which people should conform to local cultural norms. This may be particularly the case for older people, who form the largest group experiencing low income in rural areas (see, for example, NAO, 2002; Lowe and Speakman, 2006; Commission for Rural Communities, 2007a; Moffat and Scrambler, 2008; Vera-Toscano et al, 2020). This demographic group may show a stronger desire than other groups in the population to be independent and self-sufficient and may be more anxious about protecting their privacy.

Building on the importance of self-sufficiency in rural locations, evidence suggests that rather than claim benefits, rural individuals may prefer to seek a second or third job or to work informally, or prefer to live a more spartan, self-sufficient or self-reliant existence compared with urban dwellers (Sherman, 2006; Public Policy Institute for Wales, 2016). This was found to be the case for older people in research by the Commission for Rural Communities (2006). They found evidence of older people managing their finances with great care, avoiding seemingly unnecessary expenses or incurring debt and having very modest expectations in relation to their finances. Even where finances were stretched there was evidence that people sought to manage without seeking additional support from the state, relying instead on the help of informal sources. This reflected a widespread resistance to becoming dependent on the financial support of the state. More negatively, however, the research revealed the extent of change in rural communities recently, with a general deterioration in the quality of social relationships (as well as a loss of local services) perhaps threatening the availability of informal sources of help and leading to increased isolation.

As mentioned earlier, the majority of those of working age facing low incomes in rural Britain are in work and experience poverty or unemployment for relatively short spells (alongside groups such as older people and lone parents who tend to face longer-term poverty). While for some people benefits payments are vital in assisting them to cope in these periods (Chapman et al, 1998), others may choose not to claim for these times in between (for example, seasonal) jobs, given the delays and complexity involved and perhaps a fear of sanctions. Indeed, research has found no evidence of welfare dependency in rural areas; on the contrary, people are eager to find work and be independent (Phimister et al, 2000; Shucksmith, 2000; Vera-Toscano et al, 2020). As Vera-Toscano et al (2020, 226) conclude, the policy challenge is less about 'scroungers' and more about increasing claimant rates among those eligible during short periods of hardship: 'Stigmatisation of welfare benefits, tightening welfare conditionality and sanctions will reduce the effectiveness of social policy during such short spells of need'.

Another reason for lower benefit entitlement take-up in rural areas that has been suggested in previous research is that rural residents often find themselves geographically and socially apart from others in similar situations who might act as informal sources of information (see, for example, Shucksmith, 2000). This is not usually the case in urban areas where those in poverty are more likely to be concentrated in particular areas, and often in social housing. Social housing is lacking in rural areas, so poorer households are most likely to be in private housing, without social landlords to provide an effective channel for information on benefits and rights and appropriate services to reach those eligible for state support – as well as the challenge of private rented housing often being more expensive, placing an additional financial burden on rural households. Research has found that pensioners' lack of information on benefits may be due to their limited contact with service providers. They may only infrequently come into contact with GPs or district nurses who may have limited understanding of, and little time available to discuss, benefit entitlements (Bramley et al, 2000).

Accessing advice and information in distant urban centres has also been found to be problematic (see, for example, Shucksmith, 2000), particularly for those who do not have access to private transport and for whom public transport is not an option (for example, older people or those with mobility problems) or is too costly or running at inappropriate times. This may be one reason why research has suggested that rural people are more often confused about the benefits available and their entitlement to them.

Finally, it is worth noting the challenges with benefit take-up among one important group in the working population of many rural areas – migrant workers working in some of the key rural sectors, including agriculture (production and processing), hospitality and tourism. 'Seasonal' workers (who may be in the UK for a few weeks during one summer or a few months of every year, perhaps bringing with them dependent children) may lack information and understanding of the benefits to which they are entitled, resulting in under-claiming (Thomson et al, 2018; Fernández-Reino and Rienzo, 2022).

Conceptual framework for this study

This chapter has outlined the extent of rural poverty as revealed in previous studies, the factors contributing to poverty and financial hardship in rural locations and the reasons why such challenges often remain hidden. It is apparent that many of the policies acting on low-income households in rural areas are not well-informed by an understanding of the everyday experiences of living on a low income in rural contexts, and this is therefore an important element of our conceptual framework in this book.

As noted in the introduction, for various reasons, we used financial hardship and vulnerability as a prism through which to enquire into the broader, underlying processes of social exclusion in rural areas. To that end, our focus was on the interconnections between individuals' and households' everyday experiences of financial hardship/wellbeing and the structural and external processes bringing social exclusion through changes, for example, in local economies, employment, housing markets, welfare support and services, mediated through place. This necessitated analysis at both the individual/household level, enquiring about experiences and causes of financial hardship and vulnerability and revealing strategies and sources of help and support, alongside analysis of the economic, social and policy context through which processes of social exclusion operate to generate or redistribute financial hardship and vulnerability (or vice versa for processes of social inclusion). This context of processes of social exclusion surrounding financial hardship and vulnerability is presented in Figure 2.1. The processes which operate in each of these contextual elements are distinguished for analytic purposes as: markets (access according to market forces); state (access according to need or other bureaucratic criteria); voluntary and community sector (access according to charitable or associative criteria); and family and friends (access according to reciprocity).

This framework draws heavily on the earlier work of Commins (1993) who suggested that social exclusion should be defined in terms of the failure of one or more of these four systems:

1. the democratic and legal system, which promotes civic integration;
2. the labour market, which promotes economic integration;
3. the welfare state system, promoting social integration;
4. the family and community system, which promotes interpersonal integration.

For Commins (1993), one's sense of belonging in society depends on all four systems. Reimer (1998) subsequently built on Commins' work to also propose a fourfold system of social exclusion and inclusion, arguing that it is helpful to distinguish the dimensions of social exclusion according to the different means through which resources are allocated in society (for more discussion of both Commins' and Reimer's approaches, see Philip and Shucksmith, 2003). Reimer's (1998) four systems are:

1. private systems, representing market processes;
2. state systems, incorporating authority structures with bureaucratic and legal processes;
3. voluntary systems, encompassing collective action processes;
4. family and friends networks, a system associated with cultural processes.

Figure 2.1: Conceptual framework guiding the research

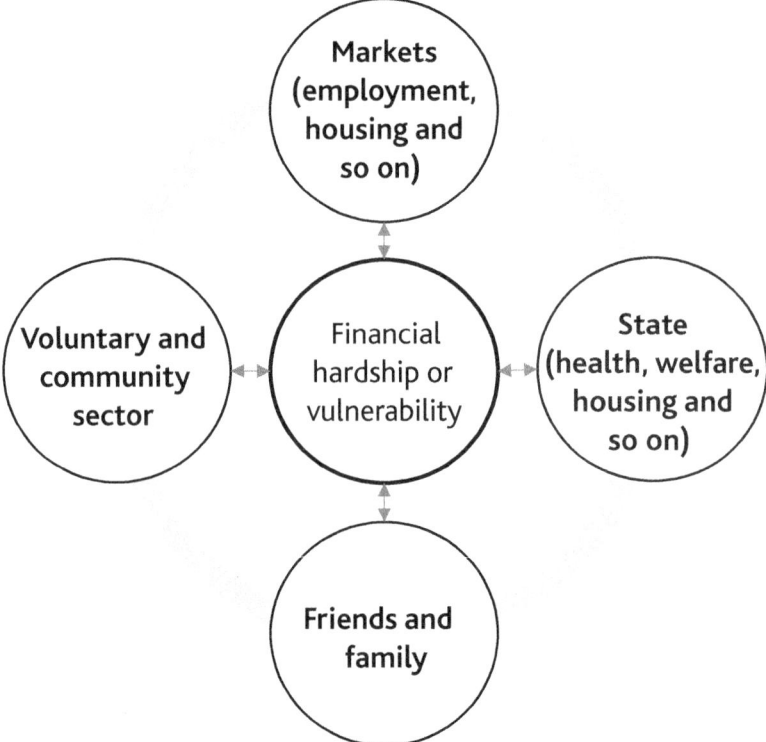

This research builds on the work of Commins (1993) and Reimer (1998) to identify the four contextual elements shown in Figure 2.1, and to explore how changes in each of these four systems of resource allocation affect individual/household financial hardship, wellbeing and vulnerability in rural Britain, through qualitative work in our three study areas and through analysis of secondary data. This focus, and the literature reviewed previously, informed the topic guide for our interviews and the subsequent thematic analysis of the interview transcripts which forms the basis of the case study chapters in this book.

Our emphasis is on exploring and understanding individual, household and community lived experiences of financial hardship and poverty, with reference to existing literature and to secondary data analysis, and the multidimensional dynamic surrounding processes which underlie their hardship or vulnerability and describing these in a thematic way. Our themes include: exploring narratives of loss, decline, place, identity, belonging and the rural idyll; exploring peoples' experiences of the (different) labour markets in the three locations and of the changing nature of work in the context of globalisation and neoliberalism; the impacts of changes to systems of state support provision; the important and evolving role of voluntary and community sector organisations and of family and friends support networks; and the impacts of the mobility of people into and out of these regions on both a permanent and temporary basis (see also Willet, 2021 for an in-depth analysis of similar themes in relation to the Cornwall region of England).

Our approach responds directly to calls made previously (see, for example, Shucksmith et al, 1996; Philip and Shucksmith, 2003) for research to move beyond statistical approaches to 'count the victims' of rural poverty, towards researching the dynamic experience of disadvantage and social exclusion in rural areas, and to understand better the processes causing disadvantage in a variety of rural contexts and their uneven impact on different groups and different areas. This can be done through detailed case study research and/or the use of longitudinal quantitative data to contextualise experiences of poverty within the local social and cultural contexts that shape peoples' everyday lives (see also Milbourne, 2004; 2014; Cloke et al, 2007).

Conclusion

This chapter has described the extent of rural poverty and disadvantage with reference to previous studies and has discussed how these challenges often remain hidden in rural places. There are many factors which have contributed to poverty and financial hardship in both rural and urban communities in recent years, including austerity policies, changing patterns of work and changing attitudes towards risk, with greater emphasis placed on the role of individuals in determining their situation, rather than on the role of the state. However, there are also a number of factors which contribute, usually in combination, to poverty and financial hardship in rural locations, including perceptions of the 'rural idyll', the traditionally strong sense of self-reliance in rural locations and a greater fear of stigma attached to claiming benefits, the higher cost of living and challenges with local service provision.

These particular circumstances, combined with the hidden nature of rural poverty, mean that policy responses have not always been well tailored to tackling this challenge, and therefore researchers have argued that there is a need for community-led responses, alongside state support to tackle structural issues. We have concluded this chapter by outlining the conceptual framework that guided our research in the three case study locations across rural Britain.

East Perthshire: an accessible rural area in Scotland

Many rural areas in Britain are within an hour's drive of a major urban centre, making urban labour markets and services accessible to at least some of their residents. Often, such accessible rural areas attract migrants both from more remote areas and from urban areas in search of 'the best of both worlds' (Shucksmith et al, 1996). The East Perthshire study area, shown in Figure 3.1, is an accessible rural area of Scotland, mostly within commuting distance of Perth, Dundee and the central belt. The area includes a population of about 19,000 people over an area of 468 km^2, half of whom live in Blairgowrie and Rattray, the principal town. To the south, fertile lowlands include smaller settlements, while to the north a series of remoter glens stretch up into the Grampian mountains. The area is renowned for the growing of soft fruit and its rich past in textile weaving. It also attracts tourists, commuters and retirement migrants, and older age groups are over-represented. The population is economically and socially diverse, including not only areas of expensive middle-class housing but also communities that are in the 20 per cent most deprived within Scotland. Accordingly, there are big disparities in the housing affordability ratio (house price/household income) from 2.3 in Rattray or Alyth to nearly 13 in the scenic north-west of the area towards Highland Perthshire.

According to Perth and Kinross Council's East Perthshire Locality Profile, which used the ACORN segmentation[1] to describe the household types within East Perthshire, the two largest groups are 'Comfortable Communities' (37 per cent) and 'Financially Stretched' (28 per cent). This indicates 'the very diverse nature of what is a relatively small and low population area' (Perth and Kinross Council, 2015a, 4). Perth and Kinross Council estimated in 2015 that 27 per cent of households fall below the Minimum Income Standard, even without taking higher rural costs of living into consideration (Perth and Kinross Council, 2015a, 48).

Figure 3.1: Map showing the East Perthshire study area

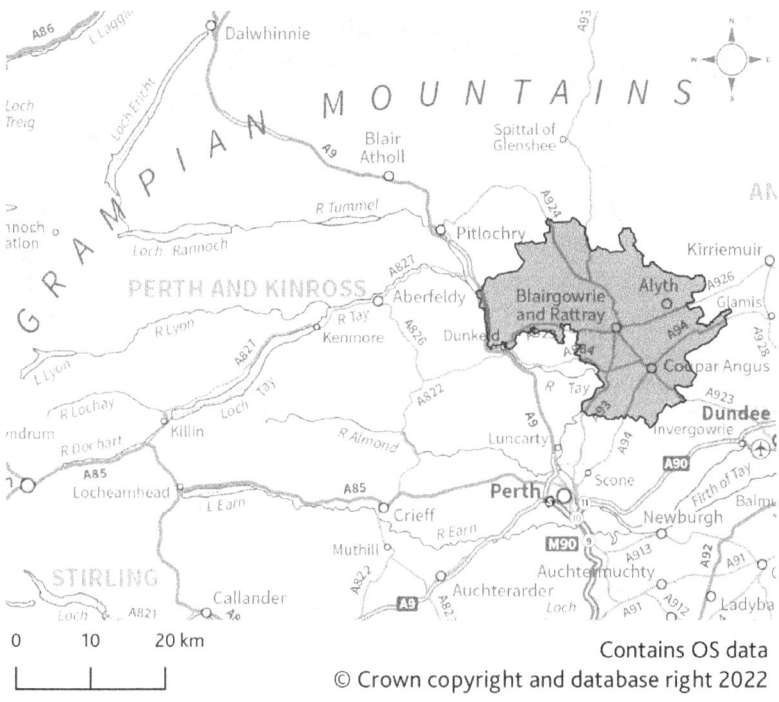

Contains OS data
© Crown copyright and database right 2022

Most people in East Perthshire live in commuting distance of cities, but the north of the study area is mountainous and remote

Income from employment

The local economy used to be based on soft fruit growing, textiles and manufacturing but fewer are now employed in these activities, with the major employers, such as Smedley's which employed between 400 and 600 staff, having downsized or closed. Although 7 per cent of residents still work in agriculture, most work in services and a growing number commute to work in Perth, Dundee or further afield, often having moved to live in this area. These commuters tend to have professional occupations and higher incomes, whereas those employed within East Perthshire are more likely to have lower-wage, less-secure occupations in manufacturing, retail, wholesale or agriculture (Perth and Kinross Council, 2015a, 45). This is bringing social changes and also affects housing opportunities, as discussed further later.

Although most people can find work, it is hard to get regular, secure, well-paid employment without commuting out of the area. The need for better-quality, better-paid employment opportunities for local people was highlighted by many groups, during the Community Planning Partnership's place-based scrutiny (Perth and Kinross Council, 2015b, 299), and they noted problems of youth unemployment. Often, the available work is insecure and unattractive – the chicken processing factory in Coupar Angus was frequently mentioned by respondents in our interviews, for example.

> 'In the rural situation that we are there are lots and lots of people who do seasonal work … but they do work that's not any guaranteed hours … If you don't have any guaranteed hours and you're in rented accommodation where you have to pay your rent each month, so life just became very, very difficult for people like that.' (Gatekeeper)

> 'Certainly here, it's seasonal work. It's low-paid work, zero-hours contracts and a lot of it is in the tourism/hospitality sector.' (Focus group)

Transport to work was a related issue, both in terms of its cost and the inadequacy of public transport. This intersected with the precarity of work, in that it is costly to travel to work for a short shift or to find you are not required.

> 'The work is available in Perth but of course if you're in precarious employment, part-time, zero-hours contracts, it can actually cost you more to get to Perth than you're actually going to earn. Because we did see one young woman, it was going to cost her about three quarters of her wages, just to get to and from Perth, to do a four-hour shift … and they might not want you at all or only want you for an hour or something.' (Gatekeeper)

The fertile land south of the study area is renowned for berry farming

Some of the unskilled, insecure, unattractive work – such as in the chicken processing factory, in berry picking or in the care sector – is undertaken by European Union (EU) citizens, some transient and some settled. Blairgowrie has a long-standing relationship with Poland since the Polish Army's 1st Tank Regiment was based here during the Second World War and most EU citizens in the area are from Poland.

> 'Lots of the people who do come from the EU to work, they're often here for the whole year now … So many of the people who do the care work, the care system, are from abroad and are really helpful … We just don't value them enough.' (Gatekeeper)

The recent growth of insecure employment or self-employment, and of low and unpredictable earnings, makes it hard to budget or to save, so increasing the risk of debt problems. Few workers are thought to enjoy work-related pension schemes. Foodbanks told us that most of their users are of working age and are in work but are still in desperate situations.

> 'We do see it [precarious, casual employment], certainly, absolutely, lots of people who have got no contract at all, no written contract and therefore are not exactly sure what their terms are … Agency workers can have less secure employment rights.' (Gatekeeper)

> 'In-work poverty is definitely on the increase. I think, as well, it's zero-hours contracts aren't helping because it's very difficult on an individual to budget not knowing. Sanctions, I think, are becoming more and more frequent in terms of the DWP [UK Department of

Work and Pensions] sanction their UC [Universal Credit] payments so they don't get what they're entitled to. I mean, with one guy they left him with £30 to live off.' (Gatekeeper)

Lack of childcare was also identified as another barrier to finding employment. Gender issues arise, as there are few employment opportunities for school hours in rural areas. This is an issue for EU migrants especially who may be less integrated into social networks and whose spouses or partners may experience social isolation, left in the home all day without transport.

'Lots of families that are unable to work because there's childcare issues, they always say, "It doesn't pay me to work", because they can only work X amount of time and quite often, that doesn't fit in with childcare.' (Focus group)

In remoter locations, in the Glens, there may only be casual, low-paid (often seasonal) work on traditional estates available, and less paternalistic care for employees than in the past. The case of one tenant in insecure accommodation who had to cycle many miles to Perth to claim Housing Benefit and visit the foodbank is considered in more detail in Chapter 7.

Self-employment was often associated with poverty. We were told that small enterprises tend to relate either to tourism, creative services or care and wellbeing. Creative and care entrepreneurs are passionate about their work and pay less attention to the returns from their businesses.

'The creatives are a good example of where I am sure many of them are living on next to nothing. They do it because that's their passion, that's what they do … The care sector is another area where low levels of income are … perpetuated by local authorities' structure, what their expectations are that a carer should earn.' (Gatekeeper)

There were very few mentions of personal savings in these interviews, except in the context of pensions and of budgeting, for example, for car insurance or children's Christmas presents. It was noted that few people in this area would have employment-related pensions "because there are a lot of small employers" who wouldn't offer such schemes, leaving it to employees to make their own provision (Gatekeeper). Many women wouldn't even receive the full state pension. A "huge time bomb" (Gatekeeper) is that so many sole operators and microbusinesses make no pension provision and will have to work beyond retirement or rely on their families to support them.

'I am becoming aware that there are quite a number of women, particularly women who maybe gave up work for family, and they are left a bit high and dry, working on into their 60s because of the state pension age being 67 and they don't have any other provision.' (Gatekeeper)

Some people are unable to work, sometimes because of mental or physical ill health, or because they have to fulfil a role as an unpaid carer. With an ageing population, and formal care at home becoming less readily available, more people of working age stay at home to act as carers (Perth and Kinross Council, 2015a); even school-age children are fulfilling this role, supported by a Young Carers project (Perth and Kinross Council, 2015a, 35). We were told that the Health and Social Care Partnership takes the view that if a family member provides care "then they don't have to worry about that one and others are prioritised" (Gatekeeper).

This precarity was exacerbated and highlighted during the COVID-19 pandemic, with the tourism, hospitality and leisure sector hardest hit by loss of income and redundancies. The UC claimant rate more than doubled in Perth and Kinross in the first two months of lockdown, with a proportionately higher impact in the council's rural areas, and 30.8 per cent of workers furloughed, compared with a Scottish average of 25 per cent (Littlejohn, 2020). Many minority community members, including EU citizens, lost their housing as well as their jobs (Focus group) and foodbank use increased dramatically. More details of the impacts of the COVID-19 pandemic in rural Britain are given in Chapter 6.

Support from the state

Apart from earnings from employment, people receive support from the state. This comes in many forms, from central and from local government, and we have already reviewed in Chapter 1 how this has changed in recent decades, with a shift from government to governance (Stoker, 1998) alongside processes of neoliberalisation (Peck and Tickell, 2002). These changes affect every aspect of state support, including education, health, housing and welfare provision, while fundamentally changing the funding context for the local authority (Perth and Kinross Council) and for community and voluntary organisations.

This section focuses specifically on receipt of welfare benefits in East Perthshire. As mentioned in the previous chapter, welfare reforms from 1999 to 2008 were effective in lifting many vulnerable people out of poverty in rural Britain (Vera-Toscano et al, 2020). Subsequent welfare reforms have sought to reduce public expenditure by limiting eligibility, introducing further conditionality, sanctions and delays and reducing real benefit levels (except for pensions), while gradually transitioning to a new system of UC (Asenova et al, 2015). Our findings reveal many ways in which these reforms create financial hardship and vulnerability for rural residents and reduce the effectiveness of welfare in supporting people in times of need.

Complexity, poor communication and digitalisation

We heard in detail, from many sources, about the difficulties for claimants of navigating the benefit system, because of its sheer complexity, the flaws in its design and the digital skills required. Registering and making a claim is itself a complicated and lengthy online process; in addition, a very full online 'journal' needs to be kept detailing evidence of job search (job applications, training, interviews and so on), and travel is required to regular assessments, perhaps in inaccessible locations.

'Providing advice to individual people about UC or just welfare benefits, in general, is so, so complex. I couldn't even begin to do it. There's a high risk attached … High risk because the risk is high in getting the wrong advice because it's so complex.' (Gatekeeper)

'Our other big quibble with UC is that the communication between the DWP and the client is very poor … if we can't understand it, then your average claimant has got no chance. Everything is done through this online journal, so (never mind IT and digital access and everything) people have to go into the journal: if they want to report a change [in circumstances], then [DWP] put the statement saying what you're going to get in the journal as well. But it doesn't really give a breakdown, so it just gives you a figure and says this month, you're going to be paid this much. So, most people haven't got a clue about how that's actually been calculated. When you ring, even when *we* ring, you can be on the phone for 30 to 40 minutes, an hour, waiting to get through to speak to somebody, to ask for clarification.' (Gatekeeper)

'Lots of people, they forget the passwords, so they can't get into the account in the first place! And trying to make sense of it, and it is complicated because when you first make your claim there's all these screens that you have to go through because they're asking you for all the information … Then you need an appointment with a work coach, people don't always realise that. So, once you've filled all this stuff in, people think oh that's it, I've done it, but you're only halfway through, you have to make the appointment, go in and provide your evidence. So, the whole thing is complicated.' (Gatekeeper)

'We find it difficult when agencies like DWP are saying it's "digital by default", everything is online. You can't go down that route, in our opinion; you have to open up other options for people. If we did that we would isolate a high percentage of our tenants who would not be able to access any services.' (Gatekeeper)

'A lot of these benefit assessments are done in Perth or Dundee. Again, Dundee is an hour and a half away by car [from X] so to get there by public transport, never mind the cost.' (Focus group)

The complexity and high risks of the benefit system are major factors in financial vulnerability and hardship in this area, and there are many elements to this. The slightest error can lead to loss of benefits, sanctions, hardship, debt, hunger and even homelessness. As discussed further later, these problems are likely to be amplified for rural claimants because poorer households in rural locations are less likely to have broadband or 3G/4G mobile signal, may not have access to a computer and are located further from sources of support and advice.

Volatility and irregularity of incomes in rural areas

Several respondents drew attention to specific problems which the welfare benefits system has in coping with irregular incomes, characteristic of many jobs in rural areas not only in farming, tourism and self-employment but increasingly widespread with zero-hour contracts, the gig economy and agency working. Inherently, these make household budgeting harder and so increase the risk of debt. On top of this, the benefits system (especially tax credits and Housing Benefit) has never kept up with volatility in earnings, often leading to overpayment of benefits which was then clawed back too rapidly for low-budget households to withstand. UC is supposed to have ironed this out by working in real time, but it doesn't because assessment is based on notification dates rather than work periods. This was an important issue in rural Perthshire.

'From a debt point of view, the uncertainty over income [from zero-hour contracts or agency work] makes it very hard to budget and it's then easy to fall into debt because this week you can afford your rent, next week you can't. [And this has an] effect on benefits ... If you've got people who are in and out of work, or their hours change all the time, what we're still seeing a lot of is people get to the end of the year and discover they've been overpaid the Tax Credit and they've got a massive Tax Credit debt ... In theory UC is supposed to do away with all that ... but under the old benefits, they used to treat income according to the period it was paid for, now they treat it as the date when they get told about it ... We have seen clients, who've been treated as having no income, even though they were working, and then the following one they're treated as having two incomes. Now the effect of that is usually that it takes you out of UC.' (Gatekeeper)

'It's the security that's missing in social security, I believe. Obviously, the cost of living is going up … [and] job insecurity … We're seeing now more people in work coming for assessment and that's because [of] zero-hour contracts, so they've got money one week and then the next week they've not got nothing. But then you can't plan; your life is near chaotic.' (Gatekeeper)

The inability of the benefits system to deal fairly with the volatility and irregularity of rural incomes was repeatedly identified as one of the most serious causes of financial hardship, not only because it makes household budgeting hard but because it increases the risk of debt and destitution.

Delays, including appeals procedures

Another consistent message from our interviewees was the harmful effect of delays in receiving benefit which seemed to them to be deliberately built into the reformed welfare system.

'The main reason that people come [to the foodbank] with their voucher now is delays in benefits and that they experience at the beginning of UC. I mean, that was just, it was just horrendous for people. And in the rural situation that we are here … there are lots and lots of people who do seasonal work [or] work that's not any guaranteed hours.' (Gatekeeper)

We were told of a professional mother and child. She had been ill with inter-uterine cancer and after chemotherapy and radiotherapy, she was unable to return to work after six months on sick pay and had therefore applied for UC but had no money for weeks due to the delays with payments.

'By the time she was coming to us, she'd actually got a job but it wasn't starting for another month. I thought that they had emergency funds that they could help somebody with … She was just sort of wiped out after six months. She assured me that she had managed, that now she was okay. But she really needed help until this job and although she said she wondered if the new job would cover all her outgoings because it wasn't full-time, but how could they put someone …? And she wasn't as upset as I was, I don't think. I just found that really quite difficult.' (Gatekeeper)

Delays and errors in UC also caused mounting rent arrears for many tenants.

'I know that's one of the issues that the council and housing associations raise at every [welfare reform strategy group] meeting, is the escalating rent arrears because of delays in payments. [And] sometimes it takes the UC a few months to actually get the housing costs right. So, we have seen claimants making their new claims to the UC, having to wait the seven weeks to get anything anyway but when they do eventually get the money, we're discovering that the housing costs haven't yet been added, sometimes children haven't been added, so they've been paid a basic UC amount and nothing extra for the children. We see that quite often. It catches up eventually but all that causes short-term instability for people of course and leads them to the rent arrears which worries, understandably, the landlords.' (Gatekeeper)

Claiming entitlements

Previous research studies have consistently found evidence of lower take-up of benefit entitlement in rural areas, for a variety of reasons noted in Chapter 2. One reason is that rural residents often find themselves geographically and socially apart from others in similar situations who might act as informal sources of information; far fewer are social housing tenants, for example. Accessing advice and information in distant urban centres may also be problematic. Furthermore, there is more visibility and less anonymity – and therefore more potential for stigmatisation – when claiming benefits in a rural context. Moreover, it is suggested that rural dwellers tend to have a stronger culture of independence and self-reliance, allied to different subjective assessments of poverty and hardship.

There are signs that resistance to claiming entitlements is greater among older people:

'Historically, older people have always been the ones that were under-claiming benefits, aren't they, and I think that's probably still true. With Pension Credit, we do see a lot of older people who are getting state pensions and they could be getting Pension Credit top-up and they're not aware of it. Some of that is down to that resistance of "well, I don't want to tell anybody about my business".' (Gatekeeper)

'People in Perth and Kinross would probably rather tell you their whole health history than talk to you about their financial situation.' (Gatekeeper)

Universal Credit roll-out and legacy transitions

In our 2019 interviews, everyone feared transitioning on to UC. All the gatekeepers from advice services who we interviewed stated that people will be worse off and at greater risk under UC, and all of the claimants were fearful of what they would lose, and of what they anticipated would be more stressful procedures.

> 'When you go from legacy to Universal Credit, most times you're going to experience a reduction in income, but if they invite you to go from legacy to Universal Credit you get transitional protection, so your money is the same until there's a change in circumstances delaying on to the route. So, there's loads of risk just now. You have a change of circumstances, that means a UC claim, it means you lose what you had previously. People did not bank on it.' (Gatekeeper)

> 'Well, since we have the Universal Credit, the standard amount per person or even for a family is not enough. Definitely, if you go to the shops and see what your grocery bill is now, when you add up everything, it's not enough.' (Focus group)

Another issue affecting those who have transferred to UC is that deductions for debts or overpayments are applied at a much steeper rate, leaving claimants with even less to live on. And the delays in receiving UC payments make such deductions much more widespread. Perth and Kinross Council staff reported almost a 25 per cent increase in the amount paid out in crisis grants and community care grants in the first six months of 2019–20 (Gatekeeper).

> 'The horrible bit is the ongoing UC payments with the deductions … So, people have amassed these overpayments over time. And yes, they were taken back under the legacy system, but they were taken back in a much more reasonable and affordable way. You move across to the UC, and it's whack and money off. So, we're seeing families coming to us since February, like larger families with huge deductions and they're left with pennies at the end of it. It's not just happening once, it's happening every month, and it's worse.' (Gatekeeper)

> 'The law is not clear that the UC advance deduction should be part of the 30 per cent [limit on deductions]. So, what you're seeing unless there's an adviser or somebody advocating on behalf of the person, you're seeing the advance coming off over and above the 30 per cent. And so when we saw that we were alarmed.' (Gatekeeper)

> 'The other problem, deductions. So, if you make your claim now, this was one of the changes the government did introduce because of the long lead-in time and people were having to wait six, seven, eight weeks to actually get any money, so you can ask for an advance payment and they did change that to make that easier. So, people get

an advanced payment but they start deducting it then, as soon as you get your first payment and they do deduct at a hefty rate. So, whatever you're entitled to, you're not getting that full amount then anyway … All these things impact on people. And it's all impacting on the people who are the most vulnerable because they're the lowest paid anyway, otherwise, by definition, they wouldn't be claiming UC. It's crazy.' (Gatekeeper)

'Last Christmas, we had nothing, not a thing. And we applied for a social fund loan, I should say. And they started taking payments back before we even got the loan. And it was a larger amount, I mean, we were £32 a week off our benefit before we even got the money in our hand. And that caused a lot of stress and we ended up paying that for, what was it, three or four months? Which in the middle of winter, with only the one benefit coming in at that point, needing gas because it's cold, needing electric, it's night, all the extra bills. And they were still taking it before anything.' (Individual)

Physical and mental illness and disability assessments

In rural Perthshire we were struck by the benefit system's treatment of clients with long-term physical or mental illness or disability, several of whom we met. We were told of needlessly frequent reassessment of those with long-term conditions, and of disregard of evidence from doctors (known in Britain as General Practitioners or GPs) and even senior medical consultants.

'You'll be familiar with disability benefits. One of the problems is the assessments that people have to go through, the fact that they feel degraded, that they feel they're not listened to, that they go for an assessment and fail and then when you're asked for a copy of the paperwork, they'll say well that doesn't bear any resemblance to what we actually discussed, what we talked about. Maybe they were only there for ten minutes or so. GP evidence gets totally disregarded, even though you would expect the GP evidence to carry more weight … And people being needlessly assessed, people with long-term conditions and they get constantly reassessed, as if they think it's miraculously going to have disappeared overnight.' (Gatekeeper)

'One of the GPs has said, "We're loathe to help out with benefits," and I'm just getting like, "Are you joking?" and getting a wee bit roused with her. "Why would you think that?," she says, "I'll tell you why," she said. "I can spend an hour writing out all this information on a Personal Independence Payment (PIP) form, send it back and they still refuse it. I'm a GP, I've told them exactly what the problem is, and they're still saying no to this person." She says, "So generally, I don't give them any information".' (Gatekeeper)

As a result of such decisions, many people unable to work lost benefits for weeks, if not months, until their appeal was eventually heard and succeeded – in the meantime having to exist without any means of support, often incurring debts at very high interest rates. We were told that since the introduction of UC, with the shift from disability living allowance to PIP, these assessments and the wrongful withdrawal of benefits had caused a noticeable increase in cases of mental ill health.

'It's blindingly obvious, don't mind me saying so, that quite a lot of the people who come have got mental health issues who, chasing them to go for interviews or if you don't go to it … you get sanctioned … I went to the DWP Job Centre Plus in Blairgowrie to speak to them about it and that experience has sort of taught me a lot. The most welcoming person, the most friendly person, was the sort of security guard that they had on the door, there to keep ne'er-do-wells out. I mean, I was absolutely appalled by the attitude that was there.' (Gatekeeper)

'My [latest] assessment was just after Christmas, January, yes. But during the run-up to that, it wasn't stressful at first but then as it got closer and closer, it got really, really bad, having to fill out the form with all the stuff. I was panicking. It was just horrendous. A lot of places are too busy … so there is nobody that helps you, which is ironic because if you're ill, I don't know how you're supposed to fill in all the forms. It's just crazy because you would think if you're ill, it should be easier for you to get, not more difficult.' (Individual)

'I just know that my PIP is due to be reviewed round about now, so I just keep expecting a letter, which is horrible.' (Individual)

Often, it seems to applicants, especially those with mental illness, that the forms include deliberate traps to trick them.

'Oh, that's just the filling out of the form that's the hard part; and the questions sound like they are trying to trick you. Depending on what the question is. It's like can you make a meal? It's like well yes, but sometimes I fall asleep when I am in the middle of cooking, but it doesn't kind of … it just asks the question can you cook a meal? It's just a lot of the questions they just give simple answers to, but it's not always a simple answer. Like the examples they give are quite simple

'… not when it comes to mental … How it comes across is that they are trying to trick you into answering yes to the questions really the answer is, if you think about it the answer is no you can't, just from the examples that are given.' (Individual)

Treatment of European Union citizens

EU citizens form a significant minority (4.4 per cent) in East Perthshire, including 3.2 per cent from Eastern Europe, most of whom are Polish (Perth and Kinross Council, 2015a). As a result of Brexit, these residents needed to apply for 'settled status' if they wish to remain. Many seemed unaware of this.

'We've got a little project to give advice on the settlement scheme, the EU settlement scheme which will enable EU citizens to stay after Brexit, assuming Brexit happens. So, we've got an adviser who works on that and they just went out to one of the farms in Angus actually, last week and they were mainly fruit pickers and they had no idea, they'd never heard of the settlement scheme, didn't know what it was, which surprised us, really surprised us. Whereas most of the people in and around Perth, at least know what it is.' (Gatekeeper)

'There's always been difficulties for EU nationals establishing their right to reside and in particular there's a gender thing there, because if you've a couple here, the woman is maybe at home looking after the children and the partner or husband works. If they split up, which happens surprisingly often, and the woman is left with the children, she has no right to reside … and we've seen quite a lot of women in those circumstances.' (Gatekeeper)

We also encountered suggestions that the DWP frontline staff discriminate against EU citizens in relation to eligibility for benefits, perhaps unintentionally through lack of training.

'The biggest difficulty we're seeing at the moment is for EU nationals because, although the DWP deny it, there appears to be the situation where if an EU national applies for UC, they are just told, "no, you're ineligible" … even though we know that they're eligible … Then it goes to an appeal [and is eventually overturned]; but that can mean that that poor person has been seven or eight months with no income while that decision is made. Very often they will have gotten into debt, they've borrowed money in order to survive. So that's one of the big problems we're seeing, and we do have quite a lot of European nationals in Perth and Kinross, about 10 per cent of our clients.' (Gatekeeper)

Perth and Kinross Council's place-based scrutiny report (Perth and Kinross Council, 2015b) notes the contribution of EU migrant workers to the area, as well as the associated social isolation and language difficulties experienced.

Benefits advice and support

There are numerous sources of help and advice in relation to welfare benefits, although there has been a tendency for centralisation and a reduction of services under financial pressure, whether provided by the state or the voluntary sector.

> 'Things were shrunk back – most of the helpful services – because the council can't afford to give them money towards it. That does seem a shame that that's happening … So, it's not helping people in rural communities really. Yes, it's difficult for [people in the landward areas]. But there are still voluntary agencies.' (Gatekeeper)

> 'The council and housing associations will help their own tenants … Other than that, nobody really. The Job Centres do have computers that people can use, so again they can go and there is a Job Centre in Blairgowrie. In theory, the staff in the Job Centre should help them, [but] we get reports from clients saying they weren't helpful at all.' (Gatekeeper)

> 'We have been really quite blown away by the fact that there are so many older people who have no idea about attendance allowance, you know, and one of the first questions that we're asking people … if somebody says, "I'm a wee bit worried about my neighbour", and we'll have a wee chat to them, she doesn't know that she's got attendance allowance, yet she's worried about her fuel costs, the house is cold, there's a leak in the roof and she didn't know she could get this money.' (Focus group)

We have been impressed by the kindness and generosity of staff and volunteers at local level, trying their best to help and support those in need in hard times and to raise awareness of benefit entitlements such as attendance allowance. Later in this chapter, we will consider further the pressures which the voluntary and community sector itself faces in providing these services.

Institutional unkindness?

A common narrative in these interviews was of an increasingly unkind welfare benefits system, with the unkindness centrally imposed from

Westminster, and of the difficulties and frustrations of inherently kind staff required to implement this system at local level.

'What would I do? Just be kind … where do you start? Be kinder in your policies. Be more compassionate. Do not treat people with contempt … I think how did the NHS [National Health Service] ever happen, how did things like that ever happen, where did that goodness go? That's been lost … and I just think what's in front of my kids, what's in front of their bairns? When's this going to end? It's just being cruel for the sake of it. Cruel to folk that have got absolutely nothing.' (Gatekeeper)

'I just think social security needs to have *security* in it. I think safety nets have gone and I think people are being treated in a really, really bad way and it's affecting children and families and goodness knows what's happening. So, I don't know, be kind in your policies and stop focusing on cutting money because, in my opinion, that's what's happening. It's like we're going to put you on UC because we've had an epiphany walking through the Gorbals or whatever happened … I think be kinder, be open in terms of policy, be braver, but that's not what will happen.' (Gatekeeper)

'The horrible bit for benefits is it's UK government policy, but we have to deliver the news, and that's horrible for [our staff] because they do have to do that.' (Gatekeeper)

'I'm not really knocking the staff in the Job Centre … The problem is, the individual staff want to do the best they can for the claimants but they are curtailed, aren't they, by the rules and regulations of what they're actually allowed to do.' (Gatekeeper)

Job Centre staff in this rural area are more sympathetic than those elsewhere, we were told, and are receiving training in relation to mental health. However, one respondent had witnessed a training session there in which Blairgowrie Job Centre staff were given targets for the number of sanctions they should impose on benefit claimants.

There is hope, among those we spoke to, that passing of greater responsibility to the Scottish government will result in a kinder system – at least in those aspects which are devolved.

'Scottish Government has, of course, put quite a lot of money into welfare reform … which is useful. The library project [Benefit Advice in Libraries, discussed later], that's funded by the Scottish government for mitigating the impact of welfare reform … They've also put money into what was called financial health checks and is now "MoneyTalk" … Their aspirations were obviously very high and the worry then is, are they going to be able to meet it? But so far, they seem to be, and I think because they're taking a very slow and steady approach and building on it, they've not rushed at it and got out of their depth. So, so far, I'm very favourably impressed – I think they're doing a good job.' (Gatekeeper)

'The new agency Social Security Scotland is it, and there's some good stuff happening. But I think the most difficult of the devolved benefits are still to happen, that's your disability benefits. So, I think that will be the proof of the pudding.' (Gatekeeper)

Social care in rural Perthshire

There is general recognition that the social care system is under great strain across the UK, and there is evidence (Wilson, 2019) that these strains are even greater in rural areas, especially for care in the community, due to greater distances, staff shortages and higher costs of provision. Our research predates DWP's December 2021 Social Care White Paper. The Perth and Kinross Health and Social Care Partnership website states that the partnership aims 'to ensure that people will receive the seamless support they need to live active, healthy and independent lives in their own homes for as long as possible'. Access to care services and distinguishing between personal care needs and social care needs, and associated charges, were identified as important issues in this area. It was suggested that withdrawal of funding for preventative services and support was leading to poorer health, and eventually to unnecessary hospitalisation.

'I think access and charges is a big issue: charging for services, and what you are charged for and what you are not charged for, is very complex, and how you are assessed, and whether that's a personal care need or whether that's a social care need, whether it's critical or substantial. It is a really … it's a complex language and it's a complex landscape … We had a whole load of people that were functioning really well with little bits of support, and preventing them becoming unwell, possibly preventing a hospital admission. That's all been taken away because they don't qualify for services and support. So that's a *big* worry, and I think we can see lots of that coming undone.' (Focus group)

It was suggested that the complexity of this system often hampers "people knowing that they have a choice and *can* have a choice about how their care and support looks" (Focus group). As this focus group participant went on to explain, "if you've not had a good assessment, a good conversation with an informed worker, then you may not know what is available, whether it is statutory, whether it is community support or whether it is family".

On the positive side, there might be advantages in the rural context in terms of a more personalised, joined-up approach from informal (unauthorised) cooperation between health and care workers.

'I think my experience in rural areas is that there may be fewer services but, actually, some of them are much more personalised and they are much more tailored … And also, that it's much more joined up, at a very local level, in terms of health and social care. So, a community nurse, on a winter's day, she might be going out to put a pain patch on somebody, and she'll phone the carer and say, "Listen, I'm here already. Do you want me to do breakfast?" So, she'll make that cross from health to social care without having a team meeting, without having to look at her terms and conditions. That's an integrated team working around support for somebody. And that's a bit of necessity, and that's maybe just a bit of rural working. So, the health and social care integration teams are much more together at an operational level. I'm not sure about the strategic level, and I'm certainly not convinced about the budgetary level … If you were to unpick that and put that up the line and look at what that cost is and who paid for it, it would all come undone.' (Focus group)

'The other side of that is, because you've got a lack of choice, you might not actually get the six hours that you need because the one care organisation doesn't have the capacity and staff to give that menu. And then if you are living in a rural area and you need 24/7 care, does that mean you have to move to a residential care setting outwith your community? Or does that mean that you need live-in care? That's all very expensive. So, you quickly do the maths, and your savings have gone.' (Focus group)

The reduction in care at home services is partly due to care in the community budgets being slimmed, as hinted at in this quote:

'Some of those problems have become more exaggerated in more recent times. Care in the community has taken a bit of a toll actually: that's care in the community without a budget to support it really.' (Gatekeeper)

It is also partly because of difficulties in recruitment of care at home staff in rural areas:

'We have difficulty providing care at home because we can't recruit staff in rural areas … and there's a concern about Brexit from that side [since many care workers are from the EU].' (Gatekeeper)

Perth and Kinross Council can fund up to four visits per day, in some cases, but care agencies are not paid for travel time between appointments such that care at home services become impossibly costly for providers.

'Sometimes the care at home costs are more than they would be in a residential care setting but you can get … I think the maximum is four visits a day or something where someone pops in, but again it's the logistics of providing that service in rural areas because someone could be travelling half an hour to make a 15 minute call and half an hour back. We only pay the agency for the visit so they have to somehow cover these travel costs. I don't know how. They are continually throwing in the towel and giving up the contracts on short notice and all sorts of things.' (Gatekeeper)

Clearly these problems in providing care services in rural areas impinge on family members' ability to take up employment and have a direct impact on household incomes.

Support from voluntary and community organisations

There is support for people in financial hardship from numerous voluntary and community organisations, some local and some national. From foodbanks to advice services, from women's refuges to care providers, these face challenges of trying to provide services across a large rural area in the context of funding pressures and rising demand for support.

'We do provide an outreach service in Blairgowrie, but of course the problem for us is that Perth and Kinross is huge and trying to provide an accessible service across it … Until April, we had a service in Rattray and a service in Blairgowrie but now we just go to Blair … And we provide advice in person but also by telephone, email and more recently by webchat.' (Gatekeeper)

The support of voluntary and community organisations is vital to many who face financial hardship or vulnerability in rural Perthshire, and they recognise and value this support, as reported later.

Blairgowrie is the main town of the East Perthshire study area

Motivations and challenges

We heard quite a lot about volunteers themselves and about their motivations. Our evidence confirms previous claims (for example, Lowe and Speakman, 2006) that voluntary and community organisations have relied increasingly on retired people with diverse skills and backgrounds, but who are all highly motivated to help others in their local communities.

> 'I think we're very reliant on our older volunteers, just because that is a stable workforce for us, they tend to stay. Our longest serving volunteer had been here for 32 years, something like that, and the older volunteers do tend to stay for longer, they're reliable, they turn up, week after week, you can really count on them … They are very reliable and a committed workforce.' (Gatekeeper)

> 'The great thing about working with volunteers is they want to be here … It's quite gruelling, I think, what we ask of our volunteers – I'm astonished anybody does it really … We expect the same standard from our volunteers as we do from our paid staff: they have to deliver advice that meets the quality standards, exactly the same. It takes about 12 months to train a volunteer.' (Gatekeeper)

This high motivation is also a feature of employees of voluntary and community organisations, as evident in all our meetings and described well in the following quote:

'Most of the paid staff, again they're here, we're not particularly well paid … people could earn far more elsewhere. So they're not here for the money, they're here because of the type of work that we're doing, so it's a brilliant place to work, brilliant team, they're just fantastic, I absolutely love my job … you go home at the end of the day and you think you've done something that was worth doing and it makes a difference to people's lives and I know from talking to others that's how they feel, that you go home feel satisfied – most days anyway, not every day [laughs].' (Gatekeeper)

Despite the strong motivation and commitment of volunteers and staff, voluntary and community organisations in the area face growing pressures of rising demand and limited capacity.

'Demand for services outstrips what we can provide, certainly. We've got core services which are funded by the local authority and then we've got a number of projects which we run to provide additional services for different things. But we do a drop-in service every morning and we do appointments in the afternoon, here in Perth, so anybody can just walk through the door in the morning and ask for advice. Then we do the appointments, we used to do appointments within the week, but now it would be at least two weeks before you would get an appointment. Then sometimes people can't wait that long, so that puts extra pressure on the drop-ins … But on some days, we have to turn people away … Even with our core funding, it's usually March before the council tell us what they're going to give us from the first of April. Perth and Kinross has always been generous to us, which helps … We've had frozen budgets, we've not had a cut since I've been here and I know others elsewhere have had their core funding cut.' (Gatekeeper)

At community level, some places were more fortunate than others in their ability to draw on resources from community assets, such as wind farms. Discussion in one focus group contrasted the resources available to community groups in Blairgowrie with those available in Aberfeldy, a village about 30 miles to the west:

'Aberfeldy has had a *load* of investment from wind farm money. So, if you kind of did an audit of the community organisations, they are pretty well resourced and established … I think that has allowed for lots of things to happen really easily – you know, like, meals going out to 70 or 80 people three times a day, there's the Timebank, there's the prescriptions, there's the shopping, there's the transport, there's the larder, there's the foodbank. And I think that that just happened because it was really resourced … When I look at Aberfeldy, the fact that it has had good community development work, good resources from the wind farm and it has a history of volunteering and a mixed community. I think that has helped to get some good care and support and to have a bit more choice for people.' (Focus group)

Help and support

Although there are numerous sources of help and support available to residents of East Perthshire, many are located at a distance in Perth. Registered social landlords will help their own tenants, and Perth Citizens Advice Bureau (CAB) will give advice by phone, email or webchat, as well as providing an outreach service in Blairgowrie cottage hospital one day a week. Poorer households in rural areas are more likely to be digitally excluded as well as being physically distant from sources of such help.

'I think that that welfare entitlement, applications for PIP and ESA [Employment Support Allowance], and actually knowing what your entitled to, they *are* bigger issues. And they are maybe more difficult to get in a rural area because there's not the independent advice and support for people … So, I think it's difficult for people to know what they are entitled and to get help to fill in the forms, and then to challenge decisions when they are not in their favour.' (Focus group)

'We did have a CAB outreach in Blairgowrie but six months ago it was stopped due to lack of funds … It was at the point that they were having two drop-ins a week and you would arrive and there would be a queue out the door. So, by the time you got to the front of the queue, a lot of the time that was their time up and they would have to ask you to come back next time.' (Focus group)

'[A lot of people] will not engage with PKC [Perth and Kinross Council] or the social work service now … It's just it's too complicated. Our system in PKC is they'd really like everybody to phone the central number and go through a list of questions and answer to see if they can get access to services. But a lot of my service users are like, "No". There is that real barrier against technology and doing it over the phone and having to fill in online forms and stuff like that.' (Focus group)

The Job Centre in Blairgowrie has computers for clients' use, but staff are not normally supposed to help them. There are computers available for use in public libraries, but these have no privacy and staff are not qualified to offer advice and support, other than when CAB staff visit public libraries to offer advice through the Scottish government-funded Benefit Advice in Libraries project.

> 'We've got a project called BAIL, Benefits Advice in Libraries, and we set it up specifically to respond to the digital need and we've got an agreement with the libraries, that we can use their computers. So, we can actually go out and sit with somebody at a computer and help them make their claim online and manage their claim online and it's great.' (Gatekeeper)

Many providers are pursuing digital approaches in responding to the challenge of reaching rural residents, while some are continuing to offer non-digital contact by telephone or through physical outreach activities, such as drop-in consultations. This plurality is important, not only because some people do not have digital access or capability, but also because people with mental health illnesses may only be able to cope with one or the other medium.

> '[We're based in Perth], but they can still phone us and that. So, we have an advice line that's Monday to Friday, you can email us, we've got a Facebook page.' (Gatekeeper)

> 'I would find [going to the CAB] really difficult, yes. I could talk to somebody in one of my groups but to go to somebody I didn't know in a place I didn't know, that's a nightmare. [Interviewer: Would emails to the CAB be easier for you?] Yes.' (Individual)

> 'I prefer talking to a person. And that's mainly, I'm dyslexic as well. So, for spelling, I feel stupid spelling with the spelling mistakes … And again, with having family, if you're on the phone, the kids tend to stay away, they know that you're busy on the phone. If you're on the laptop, you're playing, so we can make a noise, we can play about, we can get your attention. So, things take longer and the more family you have, obviously the harder it would be.' (Individual)

Despite this range of provision of support, people often had difficulty in finding appropriate help.

> 'Knowing where to find the support is quite hard … So, it's kind of the signpost; and like the GPs and that aren't always aware of places like this. I just think that not everybody is aware of the different support systems that are out here. Especially in rural areas, it's harder to reach, it's kind of harder if you're living in a rural area to find the support …

And [if you didn't have a car] there is not a bus that goes from here to there.' (Individual)

Even when they located a source of support, people found it frustrating and difficult if they found themselves speaking to a different person each time, sometimes without knowledge of the area and of the rural context, and perhaps passing them on to other departments. It was much preferred if a personal connection could be established and trust built up.

'The phone line at the Job Centre, they don't have an allocated person for an allocated area. And it could be somebody down in England and you're trying to explain what's happening up here and that's impossible because they can't picture us in such a close-knit community when they're living in such a big city.' (Individual)

'I think that lack of advocacy and support to get your benefit is [an issue], particularly in rural areas. Welfare Rights used to come out and do outreach work ... and that service was withdrawn. People are not going to go down to Perth, and not everybody is comfortable with doing it on the phone. It's quite an intimate and personal thing to do. So, I think that lots of people struggle and lots of people will not be getting their benefit.' (Focus group)

For people suffering with mental health challenges, a personal connection is vital.

'I wouldn't have talked or spoken to you six years ago ... I've been under the mental health team for the last six years, but I think coming here [Wisecraft][2] and going to the walled garden has probably, well, has made the biggest difference ... I go to another group on a Tuesday as well, that has made a difference too, but it's been ... it's not been the NHS that's made much of a difference ... I just think this fills a gap that is missing in the NHS because there's only so much they can do.' (Individual)

'I'm just exhausted all the time. I just want to sleep all the time. [Wisecraft helps] big time. I have a lot of paranoia. Coming here and speaking with everybody, you learn how things actually are and what you think they are. That helps a lot. It helps take away the paranoia for a short time and keeps you grounded. It's quite good, yes. In that way it's really, really good.' (Individual)

'Yes, [she] is really good because I had the paperwork for the fitness to work, or whatever the form is that you fill in with all the questions, and

I did the basics of it but she helped me write out all the like information and what to put in it, because you read it and you panic … it just makes you panic when you get this big form to fill in.' (Individual)

A close association between poor mental health and debt was also very apparent. Debts were referred to frequently in our interviews, either as a consequence of shortcomings of the welfare benefits system and/or as following from job loss or marital break-up.

'I think that it's well recognised that debt and poor mental wellbeing go hand in hand. It doesn't matter which one comes first, whether it's the poor mental health which then causes the debt, or the debt causes the poor mental health. As a national organisation, we would say that about a third of our clients attempt suicide or consider suicide before actually asking for help. I think locally we would say 90 per cent of our clients are either clinically depressed or have some level of stress and anxiety caused by their debt.' (Focus group)

Most notable was the role of delays built into UC pushing people into debt, the impact of waiting for incorrect assessments to be overturned on appeal and the danger of benefit overpayment and clawback arising from unpredictable incomes.

'What we have seen recently, is a big increase in priority debt and that is really worrying … Priority debt is where, if you don't pay the debt, something dire is going to happen to you. So rent, mortgage, council tax, because of course council tax have quite fearful enforcement open to them. Utility debts, where you might be at risk of being disconnected. So those are what we would call the priority debts and we've definitely seen an increase in that … Obviously if you're not paying your rent, you're likely to lose your home. So that is very worrying, that people are getting into more difficulties with those priorities and what that suggests to me is that that means people are simply struggling and cannot afford the essentials of life.' (Gatekeeper)

Respondents were asked how they coped with unexpected bills or debts and they had various strategies and sources of help. Payday loans had often been turned to, but these were universally seen as carrying considerable risks and storing up problems.

'It depends on what kind of bill it is. But we have in the past missed bills, cancelled direct debits to pay another, these kind of things, just because there's no money there … But we've never been to a payday lender. We had heard horror stories about that so we kind of avoided them.' (Individual)

Advice and support for people in debt was hard to find but, once in contact, they benefited considerably from the help and support given by StepChange (a debt advice charity), Christians Against Poverty (CAP), CABs and other voluntary and community organisations. Foodbanks were another vital source of support, and it is notable that there had been a rapid increase in recourse to foodbanks in Blairgowrie and Perth, as in other areas of the country. During the financial year 2018–19, 472 people received food from Blairgowrie foodbank, but figures for the 12 months up to September 2019 show 701 people received food there.

We have been impressed and inspired by staff and volunteers, in both the voluntary sector and in PKC, who offer support and expert advice to people experiencing financial hardship to help them keep going and overcome the difficulties they face. Apart from their kindness and compassion, they were angry about what they saw as a deliberately hostile environment which has been created around welfare rights. Some called for specific changes, such as removal of the five-week period without benefits, while others sought a more general change to reinstate kindness in the system. These sources of help are vital for many vulnerable people.

Support from friends and family

Previous research has emphasised the importance of family and friends networks in rural areas (Jentsch and Shucksmith, 2004; Black et al, 2019), both as a source of support, for example, in finding work and accommodation, and as a source of social control and stigma. Black et al (2019, 11) found an 'overwhelming and increasing reliance of young people on family for support [to fill widening gaps in social protection provision] which generated further inequalities through what might be termed "secondary impact austerity"'. Similarly, family and friends have been found to play an important role in the Blairgowrie area. The reliance on family members for care at home due to the difficulty of recruiting care workers in rural areas has already been noted.

Individuals facing mental health challenges received help not only from voluntary organisations such as Wisecraft but also crucially from members of their families, without whom they would not have been able to access those organisations' help or start to rebuild their lives.

'It's just me and my dad now … [My dad's] been really looking after me. He's getting on a bit. If it wasn't for him, I'd be really struggling. I find it really difficult to even go out of the house sometimes, especially going into shops and stuff … He does all the shopping and stuff …

We manage to get by on that [my dad's pension and my ESA] okay.' (Individual)

Among council tenants in Rattray, information and advice were shared to help maximise income from benefit entitlements, but this was less apparent in the more landward areas.

'I'm that nosey and if I think somebody's not getting something they should be getting, I tell them. I point them in the right direction … My mum was not getting pension credits … The lady across the road, we've got her an attendance allowance … There was a girl that used to live next door. She was forever at the foodbank because she couldn't make ends meet but it turned out she wasn't getting child tax credits … So, yes, I'm nosey, but it's usually for the better good. So, we've got them that. So, it's all the little things.' (Individual)

In the remoter homes of the Glens, claiming benefit entitlements might be anathema, but friends and neighbours are a vital means of everyday support in other ways, without which life and work would be very difficult.

'Even though everybody is pretty much separated by a quarter of a mile, or half a mile, or whatever, or more, there is a *really* good sense of community up and down the glen. You know, for instance, if I am working and I get held up, you just ring the [school] bus driver or a friend or whatever and just say, "Chuck my kids off at their house and we'll get them later" … And they do the same with us … Without that friendship group and community, you would not be able to live here – not with a family. It just … it would be an absolute non-starter. You need to have … because we have got no family in this area … There are quite a lot of people who are in the same boat, so everybody just helps each other out.' (Individual)

At the same time, until you are accepted into the established social groups it is hard to find work.

'You don't get work around here … not if you want to work out and about and in the hills. You don't get work until you know people. And until you've met somebody in the pub or you've gone and done some door knocking around gamekeepers' houses and that sort of thing, there just isn't work there … Initially, yeah, it was knocking on doors. When I first moved up here, for instance, I literally went and knocked on his [the head keeper's] door, and he looked at me like, "Who the heck are you?" … [Now] that has worked out well, but it has taken me … from that initial going and knocking on his door, it has taken me probably two and a half to three years to the point of now, where I am the go-to person when they need extra help – knowing you're reliable and knowing that you are physically fit to go up on the hills and be trusted.' (Individual)

In our focus groups it was suggested that family networks and support may contribute less in this area than our other study areas because other family members live far away or because the population is more transient.

'A lot of folks aren't indigenous, so the family support networks are further away. I think places like Harris, people may tend to stay more local or come back. That doesn't seem to happen so much around here, from my experience, so that could be part of that.' (Focus group)

Nevertheless, families remain important as a source of support in crisis, so long as they themselves have the resources with which to help.

'When you said that people in Blairgowrie are not asking for help from their family or relatives, I think sometimes they do not ask for help from their families simply because the family is in exactly the same situation.' (Focus group)

'We are, right across the board now, being asked the minute we meet a family to do a genogram which is to look at what family members and who is around for these children so that if, in the event they're coming into care or there's a blip and the parent can't manage and maybe needs support, we're accessing family members and we know who those family members are right away rather than reacting in a crisis. So, it is asking families to do more, grannies. We have huge levels of kinship care placements rather than foster care placements.' (Focus group)

Housing, cost of living and access to services

Housing

East Perthshire is an attractive, if diverse, rural area. It faces housing pressure both from in-migrants attracted to the area's amenity and from commuters. The Rosemount area, adjoining a famous golf course on the edge of Blairgowrie, for example, is "low density and leafy suburb-type housing … where they all commute to Perth and Dundee" (Gatekeeper). House prices are slightly higher than the Scottish average, and similar to those in Perth and Kinross as a whole, but there are big disparities in prices

Figure 3.2: Housing affordability in East Perthshire

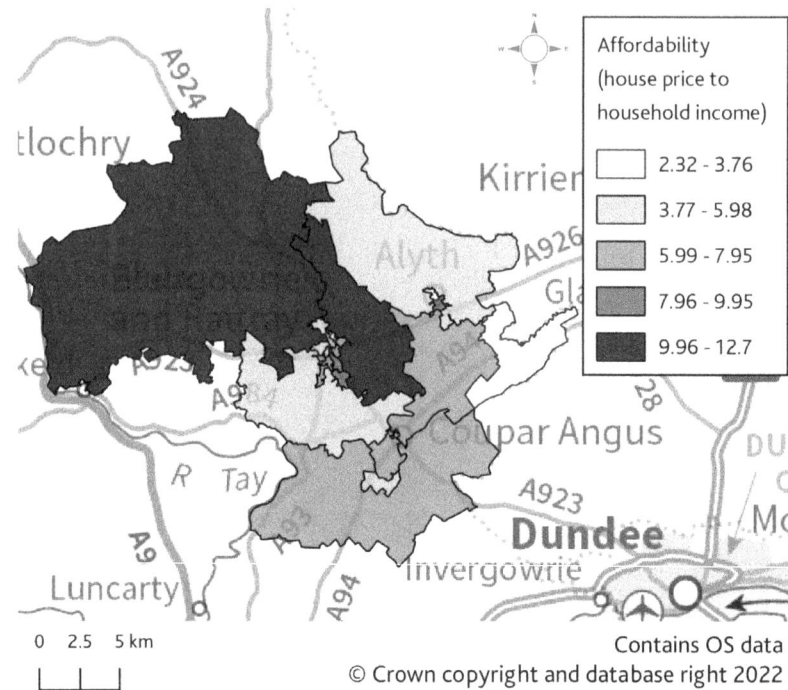

Source: Perth and Kinross Council (2015a)

and affordability within East Perthshire, with the affordability ratio (house price to household income) ranging in 2015 from 2.3 in Alyth or Rattray to nearly 13 in the north-west of the area towards Highland Perthshire, as illustrated in Figure 3.2.

The inequalities within the area both mask and stigmatise rural poverty, making it harder to address.

'I think that the wealth within the area really masks some of the poverty and underlying issues that's there, not just the rural dimension but just more broadly.' (Focus group)

'[The wealth] just hides the rural poverty which is not in a housing estate, like in Rattray, it's not in a particular street. There is very little council housing here, social housing. It's more individual families and individual households than a pocket of deprivation but it's certainly there.' (Focus group)

'I think people are not aware of how much poverty is in their area because people are embarrassed. You don't tell somebody you're getting a food parcel every week and have been doing for the last year.' (Focus group)

Income and wealth inequalities occur even within Blairgowrie, across very small distances. In parts of Coupar Angus too "you've got very high levels of poverty in one area and then you've got town houses worth £500,000" (Focus group).

'My worst benefit cases, and the level of poverty in Blairgowrie particularly, was really dramatic in comparison to all the other areas [I've previously worked in, including London], not that I haven't seen really very grim cases in the other areas … This is the first time I have seen such a dramatic difference between really, really very rich people and people just below zero.' (Focus group)

Low-cost home ownership schemes have been built by housing associations and private developers in recent years, creating home ownership opportunities at the lower end of the market. Private developers are looking to unlock further sites with affordable housing, if housing need can be evidenced, and Rural Housing Scotland (a charity which works to support the provision of affordable housing in rural areas) has been working to investigate this.

Despite new build, there has been a reduction in the number of council houses across Perth and Kinross overall, from around 17,000 in the mid-1980s to about 7,600 now, but PKC still has about 1,000 council houses in this area and these are significantly cheaper than other social housing provision. However, it has little rural provision in its future investment programme. Those we spoke to who are in council tenancies counted themselves very lucky to have secure, affordable, well-maintained homes.

Despite the low rents, about one third of PKC's tenants were in arrears in 2019, with young people most likely to be in arrears and to surrender their tenancy. The council's analysis indicates that this was a result of the introduction of UC alongside low pay and poor public transport.

'UC is a nightmare because tenants don't know from month to month how much they're going to get towards their housing costs … It's worrying that people are still struggling when the rent levels are [so low] and we are seeing genuine financial hardship, and I've been in housing 29 years … I would say the last two years it's really changed. There's people that are genuinely struggling.' (Gatekeeper)

There was some caution expressed about the growth of the private rented sector, partly because of the rising Housing Benefit bill and also because of new energy efficiency requirements from the Scottish government. There are big issues related to 'tied housing' (accommodation provided by an employer, often a farm or estate) in the Glens, which is often poorly insulated and heated (by electric storage heaters, for example) with tenants suffering intractable fuel poverty and damp, as well as facing homelessness if they lose their job. There is an oil bulk-buying scheme organised by the Mount Blair Community Development Trust (covering the communities of Strathardie and Glenshee) which reduces the cost for those who can afford oil-fired central heating.

> 'Pretty much all the houses around us would be tied or privately rented … A lot of them are in that condition, I've been in quite a few of them, you do feel when you walk in that there's a damp feeling in a lot of them. It's very old granite stone or stone-built houses. That would be across all the Glens around here.' (Gatekeeper)

> 'Just so many jobs have been attached to that accommodation and once you've lost one, you've lost the other. Even farming. I had a few cases of people that were living on farms on tied accommodation with their job. They lost their job, so they're homeless.' (Focus group)

Private tenants and owner-occupiers are often unaware of benefits for which they are eligible, and without receipt of these they are ineligible for help with fuel poverty. Many young people are said to be stuck in concealed homelessness in their parents' homes, while others are paying unaffordable rents because of their urgent need but not recognised as being in need because they are in adequate accommodation. These findings echo some of the themes emerging from the Community Planning Partnership's place-based scrutiny exercise (PKC, 2015b, 299):

> The need for more affordable and locally available housing was highlighted. … There is growth in the building of private homes, but many people reported that they could not afford these houses and there was a lack of available housing for families returning to the area. There were issues around security for families living in tied housing.

A homeowner's home is an important asset, provided they can keep up with mortgage repayments, council tax, maintenance and unexpected repairs. One respondent was unable to maintain and heat a former council house bought under the right to buy[3] in better financial times.

'It's a council house. I had some money saved up from when I was working in the hill walking shop, so I used that to buy the house, the right to buy thing. I got that cheap. So, we bought our own house … I've been unemployed since 2012 [due to mental health]. The house is falling to bits. It's still the original kitchen from the 70s. The kitchen is falling to bits. The floors are starting to go. It needs a new bathroom. It's just a mess and I can't afford to do anything.' (Individual)

A council house tenancy is also an important asset, of course, especially where (as in Scotland but not in England) the rent can be paid directly from Housing Benefit to the landlord so that residents have no worries about falling into rent arrears. This helped with budgeting more effectively.

'The benefit system pays directly into the rent. [Interviewer: And you prefer that?] Yes. That way we always know it's paid and it's paid on time. And also, we also know that the money that comes in, once we've paid the gas, electric and that, we know what we're left it without having to add, I mean, I've got two accounts, one for all the bills and one that we can use. So, that works out best for us. I'm left with direct debits coming out but the money's in the second account … We get it fortnightly, ESA's fortnightly – we work out that's best for us again.' (Individual)

It is apparent that individuals and households exhibit agency and develop coping strategies, but that their possibilities and horizons are unevenly circumscribed by the resources available to them.

Cost of living, access to services and digital exclusion

Respondents highlighted several respects in which the cost of living is higher in rural areas, including fuel costs, transport and distance to services, and higher prices.

'Of course, if you're living in a rural area, you've got all the rurality stuff as well, haven't you, more expensive for shopping and accessing services and having that bit of extra money [carer's allowance] can make a difference, in terms of being able to afford transport, maybe taxis and things like that.' (Gatekeeper)

'So, you take all that [UC] and then put it in a rural context, you've got people measuring this, saying so if your income is such and such then you're not in poverty because there's got to be our measure, and we know that the measure is flawed because it's about income. It doesn't look at what else, it doesn't look at anything else – you put that in a rural context, you could have a higher income than somebody who's living in a city, but your costs are higher, so it's meaningless, almost.' (Gatekeeper)

As noted in Chapter 2, Loughborough University's work on Minimum Income Standards in rural areas (Hirsch et al, 2013; Davis et al, 2016; 2021), a Citizens Advice Scotland study of access to food (Citizens Advice Service, 2018) and the review of rural poverty by the Scottish government (2021a) reveal in detail the additional costs of living in rural settlements. Fuel costs have consistently been found to be substantially higher in rural areas, with

Coupar Angus is one of a number of smaller service centres

consequent higher rates of fuel poverty (Scottish Government, 2021b), and this was frequently mentioned by our respondents. Some had to use expensive and inefficient night-storage heaters.

> 'By the end of the month we were running low on money because we were struggling to heat the house, because the house was quite a cold house and to have the radiator on in the living room was £50 or £60 a week in electric to use the living room here. So, we used to have to kind of top up some from somewhere [Wonga or Piggy Bank payday loans] as well … [Our council house had] old fashioned brick storage heaters – it also was a convector as well so it used a lot of electricity for that.' (Individual)

Another respondent kept the central heating switched off because of the expense and relied instead on collecting free firewood for a wood-burning stove to keep one room warm.

Fuel poverty in this area is particularly high because of the age of the housing, poor insulation, lack of mains gas and the reliance on inefficient heating sources.

> 'The properties are quite old, you're talking at least 1800s, they're not insulated very well … We're also off grid in terms of gas, so there's not gas going into properties. A lot is through storage heaters rather than oil – it's the cost of putting in. So, a lot will use solid fuel in terms of wood burners, especially up in the Glens, there's a lot of storage heating, electric heating which is not very efficient and very expensive and single-glazed windows.' (Gatekeeper)

Transport is another vital concern. Public transport is expensive and increasingly scarce beyond the main routes. We heard of people who are expected to attend appointments at the Blairgowrie Job Centre despite the lack of any public transport connection.

> 'We had one client who was living up in the Pitlochry area, so his nearest Job Centre was Blairgowrie and he was unemployed, so your basic income is £73 a week. How are you supposed to get from Pitlochry to Blairgowrie to use a computer? He didn't have a computer, he didn't have internet, he couldn't afford it. That was a while back – he got sanctioned because he'd missed an appointment and he should have been there and you're thinking there's no bus.' (Gatekeeper)

We also heard the experience of a resident in his 50s in one of the Glens, already mentioned, whose only means of accessing his nearest services in

Blairgowrie, ten miles away, was by bicycle. For the majority, as in other rural areas, access to a car is essential but expensive.

'Some of these smaller localities, some of these smaller hamlets have got no public transport. If you don't have a car then right now you can't even go to the local shop.' (Focus group)

'If your kids want to go to swimming lessons, it's a 40-mile round trip to Blairgowrie for everyone to drive, which obviously incurs costs. You know, wages *are* low, comparatively. So, the expenses of living somewhere like this are pretty high, whereas the actual wages aren't great.' (Individual)

As noted in the previous section, some people facing financial hardship were fortunate to have help from parents in buying and running a car, without which their circumstances (and mental health) would have been much worse; many others did not have this help.

'I get help with the car for like big things, like the road tax, the MOT and services and stuff. Kind of last time the car needed brakes or tyres and I went half with my mum and dad on it.' (Individual)

We also heard from many respondents that prices are significantly higher in this area than in Perth, often couched in terms of a lack of cheaper supermarkets (Lidl or Aldi). Blairgowrie's Tesco was seen as expensive (stocking premium and other brands rather than Tesco's value range), and Tesco in turn was inaccessible to those outside Blairgowrie unless they had a car.

'One of the things that a lot of the folk in the area talk about as well is that the only supermarket we have is Tesco. If you go to Forfar, which is similar in size to Blairgowrie, Forfar's got Asda, it's got B&M, it's got Home Bargains and it's got Lidl, it's got Aldi, it's got Tesco. All of those. Folk here, a lot of the women here go "I could cut £30 or £40 off the shop".' (Gatekeeper)

'I can't get into town. It's all the same to them. They don't understand that you're paying 95p for a loaf of bread at Tesco but you're paying £1.45 up here for exactly the same loaf.' (Individual)

Home deliveries generally require broadband or good mobile data access, a bank account and a fee: "you've got to have money in your bank account to do an online delivery, and you don't … some people haven't even got a bank account" (Gatekeeper). Some respondents relied mainly on charity shops, where you could buy a mixed bag of clothes for £1.

> 'We have a charity shop called [X] up here. And it's a pop-up charity shop every week. And we go in every week, usually at the end of the week, and it's fill a bag for a pound or two. So, [husband] got two black bags full of clothes today for the three of us for £4.' (Individual)

Digital exclusion is another aspect of rural disadvantage in this study area and there are several dimensions to this. One is the limited extent of broadband coverage, the slower speeds and the lack of mobile signal. Another is the cost of broadband or mobile phone contracts, which may make them unaffordable for people facing financial hardship, combined with fewer public spaces in rural areas in which it is possible to access free Wi-Fi or public computers. And, on top of these is the challenge of 'digital only' or 'digital by default' services facing those without the necessary digital skills.

> 'Digital exclusion is an issue far and wide again … I think only 50 per cent of our clients have an email address. That's probably pretty indicative of one of the issues. That's before you take into account issues with broadband and connectivity but actually just having access to those devices is a large problem … [Things are] centralised in P and K.' (Focus group)

> 'Digital access, there are still some areas without broadband and there are people that don't have the digital knowledge to be able to access the benefit system.' (Focus group)

> 'A homeless client up in the Blairgowrie area obviously didn't have a computer or any digital access and they wanted him to make this online claim for UC and then, of course, you have to manage it online and we argued, in the end the adviser went into the Job Centre with him and argued with them, that he should be able to make a paper claim.' (Gatekeeper)

> 'But the number of people who haven't got the digital skills to manage things effectively, obviously again back to older people, we do see a lot of people who have no digital access, they don't have a computer, they don't have the internet and the DWP thing is, when you go to the library, don't you because you can access it free at the library. But that is just appalling because that means somebody has got to gather up all the bits of paper and their bank statements and whatever they need and go and sit in a library and try, without the skills to actually

use the computer. A huge part of our work is filling in claim forms with people – putting it on a computer doesn't make it any easier, it makes it harder because then you've got the double whammy haven't you, of having to fill in a complex form and have the digital skills as well.' (Gatekeeper)

'In rural areas, I have to say, Wi-Fi is intermittent … Our Chair up in [X], some days he can talk, some days he can just video in sign language. Windy days he can't do nothing. So, Wi-Fi and access to Wi-Fi is a big problem in rural areas.' (Focus group)

Digital exclusion and inclusion gained further prominence during the COVID-19 pandemic lockdowns, especially in relation to home working, education, claiming government support, shopping online and social isolation, among many other aspects. These are discussed in detail in Chapter 6.

Place

Already in this chapter, it will be apparent how place interacts with other dimensions of opportunity and inequality to affect people's life chances, both positively and negatively. A frequently mentioned aspect of 'place' in rural Britain is the visibility of life in small communities, sometimes manifesting as *stigma* attached to claiming welfare benefits and deterring uptake, or manifesting positively in acts of kindness and social support. Either way, inequalities may be more visible. Both aspects are evidenced in the interviews in East Perthshire.

'If you're having financial difficulties, if you're caught in any of these traps, I think it must be so much more difficult and unacceptable to speak about it or to feel judged and that's my experience because everybody knows you.' (Gatekeeper)

'I think a lot of it is pride in that element, that your neighbour's going to know that you're not financially well off. I think that's the element. Like the foodbank, for instance, it's tucked away so no one can see who is going to it and it's very confidential and hidden. So, the stigmas are there, whether it's benefits, foodbank …' (Gatekeeper)

'We've come across various different people. What it is, it's hard to identify, because nobody wants to put themselves forward and say, "I'm struggling". So, we do struggle to get to the financial circumstances of people, apart from what they're open to tell us.' (Gatekeeper)

'I think embarrassment is part of it, shame, pride, "I should be able to sort this out by myself". Those are the usual main reasons why people don't come forward. Also, locally we find that confidentiality, because we are a small community, small community minded as well in some aspects, you've got the whole goldfish bowl thing. So, anonymity and confidentiality is something that we really emphasise about the services that we offer, whether it be for the food support or the debt advice support … [That's why] we called it a Community Larder, to get rid of that stigma of a foodbank.' (Focus group)

'One group distributed free sanitary products during the COVID-19 pandemic, with much lower take-up in the rural area than in Perth City, despite similar publicity. It's saved thousands of pounds for women and girls locally … but I do think there is something about that shame/confidentiality aspect that maybe inhibits people from coming forward and asking for help in a different way from what we see within that city centre.' (Focus group)

'I've had a lot of negative reaction, especially, I have to say, from the Blairgowrie middle classes … The prejudice! I think there is a real prejudice even in rural communities of those who have money and are comfortable and don't have to think about what it's like for the everyday life of those who actually live in poverty. They had no concept really when I was trying to explain to them. They were like, "But when are we moving these people on?" A lot of comfortable people who have had a very comfortable life, have no concept, I think, of what poverty is.' (Focus group)

Some examples were given of stigmatisation of poorer children in school.

'The things that these bairns are faced with, they've got to travel like on a quad bike, an old ramshackle thing, down a farm road to then get a minibus to the main bus that takes them into school, where you've got a clique of affluent parents that are on the parent council, probably looking down their nose at folk, getting dropped off by their Range Rovers … I think it's an equality thing. It's where it's noticeable.' (Gatekeeper)

'This young mother, she had two kids, she was a single mother, and she owed debt everywhere, and nobody could get hold of her … To cut a very long story short, she did have two jobs, her relationship had broken up, a variety of different things had happened, she didn't have a lot of support around her … [folk] were being horrible to her, all that kind of thing. So, she was thinking things like everybody's going to know because the perception was everybody talks about her, and our main concern was the roof over the head … It was just really difficult. That's in a rural area where everybody knows the mums as well.' (Gatekeeper)

Living in a small community had its positive and negative sides in the council house estate too.

'I think the fact that it's a small village but it's not too small where everybody knows everybody's business. Whereas when I was in [a smaller village] everybody kind of knew each other and knew what was going on and it was just more drama.' (Individual)

'It's been good here. Obviously, you get the same kind of problems, drug dealers, people leaving messes outside, noise. But I mean, you get these things any place. But overall, it's all right. We've been quite comfortable here. The play parks are there that our son used to play in but, come evenings, it's pre-teens and teenagers drinking, smoking, the weed, and such like, so we're kind of avoiding going out at that time.' (Individual)

'Everybody knows your business. Yes. Don't like that bit but never mind … I mean, if [we] have a fight up here at 5 o'clock, by 6 o'clock, [X] knows at the bottom of the road and found out what's going on. There's no privacy. [But] there is positives … You can trust your neighbours to taking your mail for you. Even across the road, there's an old lady across there that we go out and help and she helps us. If they're needing milk or bread or that kind of, if we're going, we'll grab their shopping list and head out for them and if we're unwell, they'll do the same for us. So, it's quite friendly, providing you pick the right one obviously.' (Individual)

In some small communities, individuals would not claim welfare benefits on principle, devoting their energies instead to finding sources of casual work and maintaining the personal networks necessary to this, while reducing their expenditure by turning off central heating and self-provisioning. These localised cultures and dispositions have been characterised as 'place habitus' by Black et al (2019), an aspect which will be discussed further in Chapter 7.

For many of the most vulnerable, for example, people with mental health challenges, just surviving from one day to the next is their priority (and time horizon), and the obstacles to progress may appear insuperable.

'It's been difficult this last while because I'm still sorting out my illness so I've just been concentrating on that. But it seems to be never-ending. There doesn't seem to be any solution to it. It just keeps going on and on.' (Individual)

'[To return to my previous occupation eventually] I would have to do a return to practice course. I don't know how I would be able to afford to, plus it means I would have to pay my rent and stuff, and my benefits would change or stop or, I don't know, I would need to have a proper look … At the minute I think it says on the letter that I can't do college courses or anything without telling them, and if I do volunteering I have to tell them. So, it's scary to change things because you don't know what it's going to affect … By the time you add up what you're getting all your benefits to what you would get on minimum wage, it's a big difference. I know it sounds bad saying it's better on benefits, but you don't *want* to be on benefits.' (Individual)

As detailed earlier, our respondents found help, support and advice from various sources, but they often recounted how difficult it had been for them to find these sources. Others had got to know welfare rights experts, in the council or voluntary organisations, and from this had developed considerable expertise themselves in the welfare benefit system to maximise their take-up.

Conclusion

For most people, East Perthshire is a beautiful and attractive place in which to live but clearly this varies according to individual circumstances. Often, it is the perspective of those with greater power and voice which predominates, and it is important to hear and reflect on the perspectives of less fortunate residents whose voices are often unnoticed. As in many accessible rural areas, inequality exists between residents with higher incomes and secure employment, often derived from commuting to nearby cities or retirement from such employment, and those employed locally who are more likely to have lower-paid, less secure occupations. In-work poverty was increasing, and a common theme in the interviews was the recent growth of insecure employment or self-employment, making it hard to budget or save.

This growth of precarity in labour markets was exacerbated, rather than ameliorated, by increasing precarity in state welfare systems. The complexity and high risks of the benefit system were found to be major factors in financial vulnerability and hardship in this area, and there are many elements to this. The slightest error can lead to loss of benefits, sanctions, hardship, debt, hunger and even homelessness. We were told in detail, from many sources, about the difficulties for claimants in navigating the system, the flaws in its design, its inability to handle irregular and volatile rural incomes, the digital and literacy skills required and the travel required to regular assessments, perhaps in inaccessible locations. These defects were magnified for clients with long-standing physical or mental illness or disability. This chapter has also highlighted the impact of delays and appeal procedures, the reluctance of some rural households to claim benefit, the treatment of EU citizens (who formed a significant minority in this rural area) and difficulties many people face in accessing benefits advice and support. A common narrative in these interviews was of an increasingly unkind welfare benefits system imposed from Westminster (institutional unkindness), and of the frustrations of inherently kind staff required to implement this system at local level. Within the constraints which austerity and welfare reforms have imposed, PKC is perceived by our respondents as having continued as best it can to provide support for those in need, and for the voluntary and community groups such as the CAB which work alongside them.

The support from numerous voluntary and community organisations, some local and some national, was vitally important. These provided a 'first port of call' and a crucial signposting role towards other sources of help. They all face challenges of trying to provide services across a large rural area in the context of funding pressures and rising demand and need, with tendencies towards centralisation and digitalisation of services and less face-to-face outreach. Support also came from family and friends networks, especially in substituting for the state's social care, childcare and eldercare services. But in our focus groups it was suggested that family networks and support may contribute less in this study area than in other places because family members live far away or because the population is more transient. There was also evidence of social stigma attaching to poverty and benefits, manifesting in different ways in different places. Nevertheless, families remain important as a source of support in a crisis, if they themselves have the resources with which to help.

Harris: an island area of Scotland

Harris lies in the Outer Hebrides chain of islands off the west coast of Scotland. Although physically connected to the Isle of Lewis, Harris and Lewis are usually considered as two islands separated by mountains. Harris is one of 94 inhabited islands in Scotland. In contrast to Perthshire, Harris is sparsely populated and has a total population of around 2,000. The main town, Tarbert (population 550), is in the north, and the remaining population lives in crofting townships around the island's perimeter (see Figure 4.1). Much of the island's interior is uninhabited. A larger town, Stornoway, in neighbouring Lewis, is the main service centre for the islands. It is an hour's bus ride from Tarbert, and even further from other settlements in South Harris. Connections to the mainland are via ferry (one hour, 45 minutes from Tarbert) or plane (from Stornoway).

The population of Harris has declined by almost 50 per cent since 1951. In 2018 32 per cent of residents were aged over 65, and approximately 41 per cent of the population resided in one-person households. Primary industries are tourism, with some fishing and crofting,[1] but as in other rural areas, most employment today is in the service or public sector, notably health and education. In the Outer Hebrides as a whole, public sector employment accounts for 38.6 per cent of all employment, compared with 24.7 per cent for Scotland, and 18.4 per cent for Britain (Outer Hebrides Community Planning Partnership, 2019, 4).

A notable feature of Harris (and the Western Isles more generally) has been localisation of governance and control. Until 1975 Harris was part of the county of Ross-shire, administered from Dingwall on the east coast of mainland Scotland, but local government reorganisation in 1975 created Comhairle nan Eilean Siar (the Western Isles Council) with its headquarters in Stornoway. Since the 1990s, control and governance has become even more local, with the transfer of large tracts of land to the community: 70 per cent of the Western Isles is now in community ownership. In Harris, most of the island is under community ownership by the North Harris Trust and the West Harris Trust. In the Bays area to the east, another community buyout is under consideration. Nevertheless, there is still a sense of distance and remoteness from centres of power in Edinburgh, London and multinational boardrooms around the globe.

Figure 4.1: Map showing the Harris study area

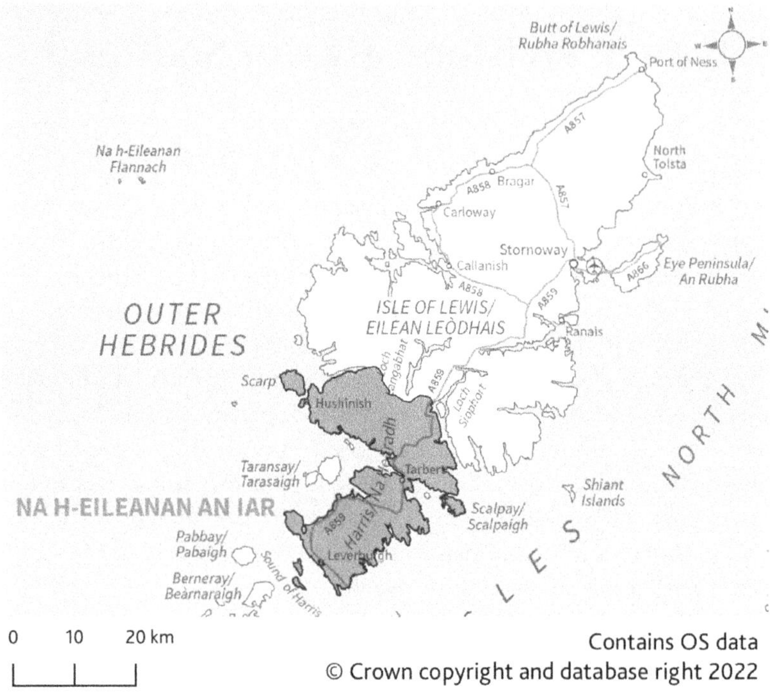

0 10 20 km

Contains OS data
© Crown copyright and database right 2022

As an overall reflection, none of the individuals who we interviewed in Harris faced the depth of financial hardship or vulnerability which we discovered in Perthshire, although we heard plenty of evidence from others about hidden poverty, especially among older people. Neither did we hear as much about reliance on state benefits. Yet, there were very real issues identified in Harris and these are considered in this chapter.

Income from employment

Like in Perthshire, there have been many changes in recent decades in the rural economy, but this has played out differently in Harris. Research in 1993 found Harris 'at a crossroads' or 'on a cliff edge': poverty, unemployment and loss of young people meant residents saw change looming (Shucksmith et al, 1996). Quality employment was scarce, and there were limited jobs of any kind. The most striking impression from our interviews is that the Harris economy has started a process of transformation. The last five years in particular have seen significant developments: a considerable increase in 'destination tourism', and major new employers in Tarbert such as the

Ferries connect Harris to the mainland, from Tarbert to Skye and (here) from Stornoway in Lewis to Ullapool

Harris Distillery and the Essence of Harris. This has put communities under pressure, particularly housing pressure, leading to difficulty in attracting or retaining staff. We were told "there used to be no jobs: now there are no people to do the jobs" (Gatekeeper).

> 'I have heard that fish farms are struggling to get employees locally, but I think part of the reason for that is there aren't that many employable people who are unemployed.' (Gatekeeper)

Tourism is clearly an economic driver but so are the community trusts, as a result of their role in job creation, housing development and the provision of small business/retail units. An increase in community confidence as a result of the trusts was also said to have had economic spin-offs.

There was optimism about further future economic potential in terms of marine tourism and renewable energy (if interconnector capacity to the national grid on the mainland is increased). As a result, Harris is now perceived by gatekeepers as having little poverty in comparison with other parts of the Outer Hebrides, although all acknowledged that there were isolated cases of financial hardship. And while Harris was seen to be 'booming', the benefits were not evenly spread, and it was still very precarious.

'The opportunities are there but we're struggling for workforce. We're struggling for people to do the jobs and we're struggling for accommodation.' (Gatekeeper)

'[Many years ago] when I started, we were looking for jobs, jobs that was what we were looking for. Now, we're looking for the people to do the jobs, and that's the way things have swung. It's a huge turnaround.' (Gatekeeper)

Housing was a recurrent theme in the interviews, in different contexts and with the link to employment being particularly important. The seasonality of employment adds further pressure on the housing market. We were told that in the past, seasonal positions were filled by young islanders – school pupils and students returning home for the summer. Now, there are not enough young people to fill the available positions so seasonal staff come from off-island, but getting accommodation is very difficult as this is also the peak tourist season. One hotel owner reported that previously he would have housed summer workers by relying on known contacts with available housing, but this is no longer sufficient so he has started to buy houses for staff use, as well as using caravans. In 2019 one café was forced to temporarily close mid-season because the chef had left due to housing issues.

'There is no shortage of work but he can't take folk on. He wants to take more folk on but there is nowhere for them [to stay].' (Gatekeeper)

'One of the biggest concerns when we were advertising for [X's] post, I was seeing the application forms coming in and I was panicking because a large percentage of them were from folk off the island. I thought, "If you guys haven't got somewhere to live, we can't even think of offering you a job because where are you going to stay?" That is one of the big inhibitors to employment.' (Gatekeeper)

For some, the lack of flexible childcare prevented individuals from taking up available employment opportunities. Private childcare (outwith the school nursery) is expensive, and many jobs that were available were said to be low paid, making this type of childcare unaffordable. Respondents were delighted to have a school nursery, which is fully subscribed, but it often does not give enough flexibility in provision to fit with irregular working patterns.

The cost of transport was said to be an issue for shift workers, particularly in the care sector where public transport was not an option, meaning they were reliant on car travel. So, while having access to a car was an essential requirement to take up these positions, these are not highly paid jobs, and this added to the difficulty in recruiting people to these posts.

A key issue raised was the lack of choice of employer: most sectors are present, but often only one employer in any particular industry. Equally, there is often only one or two of any kind of position available. This marks the employment market out as being very different to mainland and urban areas, where there would be greater choice of employer, and far greater opportunities for career progression. The only significant commuting option, to access a wider range of employment opportunities, is to travel to Stornoway in Lewis, or the mainland if the job makes this possible (for instance, offshore work, which was a fairly common strategy).

This limited range of employers, and a preponderance of small to medium enterprises (SMEs), limits the potential for individuals to progress up the career ladder, but at the same time, it means that employees get a range of experience that they might not if they were part of a larger team. People are more likely to be generalists to fulfil a range of needs within a small business, which can create business vulnerability through over-reliance on one member of staff.

> 'If you have an important member of staff they can't be replaced easily. The employment problem here isn't the lack of jobs. It's the lack of possibility and mobility assurance.' (Individual)

Although there are more employment opportunities now, the jobs are not necessarily well-paid or professional jobs, resulting in poverty in work and then beyond into retirement:

> 'You've got people who have managed to have savings but then again, there are a lot of people here because of the poverty in the past from the fishing and crofting and the things that they did, didn't really build up any savings so they're in that very poor state – its hidden.' (Gatekeeper)

As there is usually a need to travel to work, there is a financial impact due to the necessity of having a car, higher fuel costs and often long distances travelled. The housing market conditions (as well as family and crofting links to a particular township) make it very difficult to move house to be closer to a place of employment.

> 'I mean, there's not a lot of jobs; there's no solicitors and lawyers in Harris. But high-end jobs in Harris, there isn't a lot – most people have to travel to Lewis for them.' (Individual)

> 'there's nothing for pharmacists or engineers or bankers … there's decent jobs with the health service, there's decent jobs with the council, there's care home facilities and so on in the care provision … but a

lot of these are at the lower-paid end as well rather than higher-level opportunities for folk.' (Gatekeeper)

A lack of professional jobs was also reported as having made it difficult to attract young people that have left Harris for university to come back to live. Yet attracting young people to the island, either returnees or not, was seen as crucial to the future sustainability of the island. There is an urgent need to address the demographic imbalance, and while there are signs that this is starting to happen, it is still very fragile. The ScotGrad programme, which can provide a subsidised project-based graduate placement for up to a year, has been an effective way to attract young people to Harris who might not have otherwise come. Established in 2013, the programme is operated across the Highlands and Islands by Highlands and Islands Enterprise and is part-funded by the European Regional Development Fund. Local businesses have made use of the scheme and it has resulted in a small but significant number of young people coming to live and work in Harris, often staying beyond the length of their placement. In most cases, they encountered difficulty in accessing affordable housing.

> 'Yes, there might be better jobs now; there's still not enough high-level jobs. If you're a young person who has gone off to university in the mainland and you might think: "I don't want to move home because my career will just stagnate and there's not a huge amount of stuff to do".' (Gatekeeper)

It can be difficult to fill some professional posts in Harris, unless there is already someone locally with the necessary skills. And in several instances, we were told that while it might be possible to recruit one person to fill a post, the person may not come because of the difficulty finding an appropriate position for their partner, due to limited professional options.

> 'I can think of people who are living on the island right now who have moved here because their partner's got a good job here, they want to make a life here with their family and that's kids that are going to the schools and that's keeping the schools open and so forth, but their partner's not found a job yet that's appropriate.' (Gatekeeper)

A related issue is that the low level of demand for some services negates the need for creating a full-time post. Attracting a professional to a part-time position can be challenging. We discuss later in this chapter the tendency for people to have multiple jobs, which is a common response of rural residents to ensure a decent income where full-time positions are limited. For professional roles, it can be challenging to make up hours with similar work to create a full-time equivalent position. This can impact on rural service provision, with some key services either being under-staffed, delivered from a distance or delivered in a somewhat piecemeal way.

> 'they maybe don't need a full-time position because the demand is not there for a full-time position but trying to fill those positions that aren't maybe full-time has always been a big problem in the smaller communities.' (Gatekeeper)

We have already noted the increase in tourism. The 'staycation effect' of the COVID-19 lockdowns has increased tourism in Harris even further: it is typically less congested than other places and so seen as more 'pure' and 'COVID-safe' (we discuss this more in Chapter 6). Pressure on ferries in the summers of 2020 and 2021 was higher than normal and for longer than in a typical season. The power of social media is seen as being key to fuelling this change. Around 50 per cent of the income into Harris is connected to tourism, with an estimated £180m coming to the island through this sector each year (not all of which remains on-island). Such a heavy reliance on one source of income was not seen as sustainable. There was a desire to reduce this reliance by increasing the amount that comes from other sources:

> 'I don't think that's a healthy position on the economy. I just worry about recommendations that actually reduce that bit rather than bring other pieces up. I like policies that bring things up rather than try and cram tourism down.' (Focus group)

There was seen to be further potential to increase the economic impact of tourism and build a broader economic base on the back of the strong Harris brand that has developed over recent years. Visitors to the islands were said to be relatively well off and have more disposable income than the average local. Islanders were thought to be slightly 'shy' about charging high prices for things. Food was seen as an area for development and greater income generation. Harris Development Limited (HDL)[2] and others are developing a number of initiatives to extend the 'Harris brand' for food products such as lamb, shellfish and even seaweed.

Harris is seen as being rich in natural resources, and there is increasing recognition and promotion of the 'Harris brand' which is built on the natural beauty and resources of the island. This has opened up, and been driven by, opportunities to build a product base based on the Harris environment and image. The recently developed Harris Distillery makes gin that is infused with a local seaweed, sugar kelp and local botanicals, and is presented in an iconic bottle which encapsulates the designers' impressions of the 'island, its elements and people'. Similarly, the Essence of Harris company has incorporated the images, place names and 'scents' of the island into its products, so each candle 'captures' a bit of the island. Harris Tweed, which has had a recent resurgence, is a further product that has very much been associated with the island for many years. There is potential for other businesses to be developed on the back of the natural resources: as one person noted, "Harris is rich in … natural resources" (Gatekeeper).

In a labour market dependent on a few sources of employment, the economy is very vulnerable to broader changes in the wider UK and global economies, with changes such as Brexit and now COVID-19 potentially having a dramatic effect. The Brexit Vulnerabilities Index, which has been by developed by the Scottish government (2019, 3), confirms that the islands are some of the most vulnerable areas of Scotland to the impact of Brexit. All of the data zones (small area statistics areas) in Harris are in the most vulnerable category of the Index:

> Many of the areas most vulnerable to Brexit are in rural locations, in particular on the Scottish islands. Around half of communities in Shetland Islands, Na h-Eileanan Siar, Argyll and Bute and Dumfries and Galloway are amongst the most vulnerable communities in Scotland (20 per cent most vulnerable datazones). (Scottish Government, 2019, 3)

An area such as Harris is also vulnerable to the disappearance of a major employer: "[our] MP said that for every job we lose here, it's like 1,400 jobs going in a big city because of the socio-economic and these were well-paid jobs as well" (Gatekeeper). By the same token, the recent Harris Distillery development has had a huge positive impact on employment: in March 2019 the distillery was employing 30 full-time, six part-time and two seasonal staff, after only three years of operation. While in an urban context these numbers would not be significant, for a community like Harris, they are very significant.

Historically, crofting was a significant activity for people, providing an additional income from sheep, as well as other resources such as peat for fuel, meat, eggs, vegetables and crops. While there has been increased diversification and change and prosperity in the wider economy, crofting has declined. There are fewer people keeping fewer animals, ageing crofters, absentee crofters and continuing marketisation. The Crofting Commission was said to lack sufficient resources to enforce the regulations.

> 'The crofting is definitely not the way it used to be. There's some people hanging on to the crofts especially on the west side … but definitely not as much, well nothing like it was really.' (Individual)

We have noted previously that having multiple jobs is common in rural areas, Harris being no exception. Crofters have traditionally had a range of income sources, from sheep, to tweed to fishing. Today, crofting itself may be less significant an activity than previously but having a mix of employment persists. For some, this means doing one thing in the summer, and another in winter. For others, it means doing multiple jobs year-round to generate a decent income.

> 'Some people, if it's just manual work, like shop work or whatever, classroom assistant, they might have two or three jobs. I know some classroom assistants, because they've finished at 4 o'clock and they've got long school holidays. … They've got a second job.' (Individual)

> 'That's the way you live up here sometimes. You're doing a little bit of this and a bit of that.' (Individual)

Self-employment is key to the mix of work that people engage in, partly because of its flexibility to fit round caring responsibilities and/or other work, and partly because running a small enterprise from home doesn't require transport. We heard stories of people using redundancy payments to buy a boat to run tours, or to put up camping pods on their land for tourist accommodation. Self-employment comes with drawbacks, however, just as it does anywhere. One major issue in a housing market that is dominated by owner occupation is the difficulty of getting a mortgage.

Running a business from home, or even being a home worker for an employer, requires good connectivity. In a previous study (Shucksmith et al, 1996), the internet was seen as having real potential for home working and opening up the islands to more professionals who could choose to live there while carrying on their mainland jobs. While the digital connectivity of the islands has increased dramatically over the last 30 years, it still lags behind the mainland in terms of technology, capacity and speed, because of the cost of installation and the rapid advances that have been made generally. We were told there are still areas of Harris that are not served by fibre, and have very poor, and expensive, broadband as a result. Examples were given of households that had relocated, or wanted to, but had been limited by the need to secure a reliable internet connection.

Home working, whether it be a full-time job that is either island- or mainland-based, or small-scale self-employment, was a strategy for people with caring responsibilities. Working from home was said to give carers the flexibility they needed to combine care with work. In an urban area, there is likely to be a wider range of care facilities that can be accessed, compared with a rural area.

It was notable, and surprising, that insecure employment and precariatisation were said to be diminishing rather than increasing (as elsewhere). Seasonality of employment remains but the tourist season is now longer, and seasonality has always been a feature of work in the islands.

'I'm not so aware of that [zero-hour contracts]. I mean, that's a problem on the mainland … but I'm not so aware of that here and I think that's because employers are struggling to find workers. They work through the summer, the summer finishes, the hotel closes or it cuts back … they're only taking on a seasonal staff.' (Gatekeeper)

Seasonality of employment was also reported to add strain on relationships, because of the intensity of working patterns during the season. Equally, long periods at home with no work over winter could cause domestic problems. Unless additional work can be found over the winter to supplement incomes, seasonal work also requires a high level of financial management and budgeting to ensure that the relative largesse of the summer isn't used up before the start of the next season. HDL is keen to support development across the island that helps to extend the tourist season to be almost year-round, thereby extending tourism's economic benefit. The ferry capacity puts a natural cap on how many people can visit Harris in the summer, so extending the season is a natural way to increase tourism income.

Support from the state

Because of the improved labour market conditions and in-migration of wealthier households, average income levels in Harris have increased but may now be more polarised than previously. According to several respondents, those most likely to be in poverty are indigenous older people, often living remotely and reliant on a state pension, unlike people of a similar age that have moved in:

'The younger folk you see coming in with a bit of private pension … there are a lot more with private pensions now than when I started 30 years ago.' (Gatekeeper)

Those without private pensions, and others in financial need, may receive support from the state. There was also anecdotal evidence that older people were trying to cash in their employee pensions, to release income in the short term. Advice agencies were trying to caution against this because of early encashment penalties, and so they had developed a specialist advice service:

'We've got a financial inclusion officer as well, so she goes out and visits people and try and help them to get access to all the benefits that they're entitled to. Like I said, our biggest client group is probably elderly.' (Gatekeeper)

Earlier research in Harris (Shucksmith et al, 1996) reported a reluctance of residents to claim benefits they were entitled to, and there were differing views today about Harris people's willingness to claim benefits or to seek help and advice.

> 'It's a lot more close-knit community, so if people are feeling self-conscious about applying for a grant or going to a foodbank or seeking out help in some way, there is that additional barrier that might be there.' (Gatekeeper)

Apart from this, many of the issues around the benefits system found in our Perthshire study area were evident in Harris too, but there were some additional aspects, most notably the lack of assessment facilities for islanders with long-term illness or disabilities (discussed in more detail later).

Respondents in Harris spoke less than in Perthshire about the challenges of engaging with a complex benefits system and the digital infrastructure and skills required. It was suggested that form filling was such a part of island life (because of crofting) that people didn't think to mention it. Nevertheless, barriers exist: the lack of availability of IT equipment among some groups, and poor broadband, mean that these were real issues, and the Citizens Advice Bureau (CAB) has been working with others to try and distribute second-hand IT equipment to those in need in order to improve digital accessibility.

A lack of awareness of benefit entitlements was apparent, along with a need for advice and support in accessing these sources of income. The CAB network across Scotland has introduced a financial health check process, which goes through people's incomes and expenditure, to make sure they get what they are entitled to and don't pay more than they should for goods and services. The average additional income that a household benefits from, as a result of this process, was reported to be higher in the Western Isles than nationally, by 7 per cent or more, mainly due to an under-claiming of benefits:

> 'Nationally, we're averaging £2,800 benefit … to everyone who we do this financial health check on. Our figures are slightly [higher here at] about £3,000. We've had them sometimes up to £25,000. So, this might be a tax rebate that somebody is due because they've never claimed what they should have. A lot of it is benefit.' (Focus group)

> 'I think it's because they don't know about it … I mentioned it to one lady, "You should apply for attendance allowance". "What's that?"' (Individual)

As in Perthshire, the volatility, irregularity and seasonality of incomes in Harris, and the difficulties which these present for the welfare benefits system, were frequently mentioned. Inherently, these made household budgeting harder and so increased the risk of debt and rent arrears.

'We had a woman in [the foodbank] once … and she came in and she said to me it was either eat or pay the rent. And she was working but it was one of these where she would go in and they'd say, "We only need you for an hour today".' (Gatekeeper)

'Yes and she got paid early at Christmas, so again that's another period where she got two wages and then she had nothing in January to pay her rent and she's now got herself … she's had to take out a credit card, she's never had a credit card in her life, she's about 38 years old, never had a credit card in her life, she had to get a credit card to buy Christmas presents for the children … she's done everything by the book and she's been forced into a situation now because of the way that her rent is paid to HHP [Hebridean Housing Partnership] under the dates that their rent is due. Her Universal Credit doesn't go in on time and then because of her pay dates and it's just a mess of absolutely no fault of hers, so she's had to stop work – she's given up her job.' (Gatekeeper)

On the other hand, one respondent maintained that difficulties with overpayment and clawback were a feature of the legacy system which Universal Credit (UC) had resolved:

'That's one positive thing about Universal Credit. I'm not going to try to run down UC all the time, but positively what UC has brought for customers is less overpayments than tax credits. Tax credits were always renowned for huge overpayments.' (Gatekeeper)

The weight of evidence, however, in both Harris and in Perthshire, is that this remains a problem. A further issue highlighted was to do with rent payments, which can be made directly to the landlord or given to the tenant. The latter option runs the risk of the tenant spending the money rather than paying their rent, but either way, small changes in payment periods and income levels can lead to rent arrears being accrued.

'Because of the way that the rent is calculated, compared with the payment date, because it's four weekly and Universal Credit is paid monthly, there's actually 13 [HHP] rental periods in a year, whereas Universal Credit only makes 12 payments. So, there will always be one month every year where the rent doesn't get paid for the tenant if it goes directly to the landlord.' (Gatekeeper)

'Of course, the temptation is to use that money [rent element of UC]. If you're desperate and your children have no shoes and you haven't paid your rent. I mean, your children come first.' (Gatekeeper)

Notwithstanding these difficulties with irregular incomes and payment dates, tenants generally preferred to have their rents paid directly to their landlord each month because this made it easier for them to manage their budgets and to avoid rent arrears.

An issue mentioned frequently, which has a distinctive island dimension, was the benefit system's treatment of people with long-standing physical/mental illness or disability. They are required to undergo a Work Capability Assessment (WCA), not from their General Practitioner (GP) but from a UK Department of Work and Pensions (DWP) contractor (Atos). There is no assessment facility in the Western Isles and instead assessors visit the islands once or twice a year. Consequently, there is usually a long delay before people in the islands can have their WCA and receive any benefits due, during which they must continue to demonstrate their availability for work, and/or risk losing their Motability car.

> 'The other problem we have is … people claiming ESA [Employment Support Allowance] or limited capability for work under Universal Credit, we are waiting now over a year for them to get the capability assessments. I've had several clients that have been waiting over two years for Work Capability Assessment. [In the meantime] they just get the basic rate.' (Gatekeeper)

> 'Also … we transferred from DLA [Disability Living Allowance] to PIP [Personal Independence Payments], people are losing their Motability cars and of course they don't have the public transport that they have on the mainland and that's causing social isolation … So they're stranded in their homes now.' (Gatekeeper)

Corroborating this, Angus MacNeil MP's website[3] confirms that he wrote to Atos Healthcare after a constituent was left without benefit because Atos failed to arrange a WCA for over a year after applying for ESA. Mr MacNeil wrote: 'This is a shocking example of the system failing. Atos has claimed that the delay in carrying out the Work Capability Assessment is due to our remote location'. He reported the issue to the UK Minister Iain Duncan Smith MP. This was not an isolated incident, though, and the issue seems to persist still in 2020. Even when a claimant eventually has an assessment, national data show that they are often wrongly assessed as fit for work and then have to appeal the decision. Although more than half of such appeals are successful, claimants are left without their benefit entitlements during this period.

Delays in receipt of welfare benefits were a major problem for those requiring help from the state in times of need, especially for those applying for UC with the routine delay of at least five weeks before the first payment. At the Stornoway foodbank, "benefit delays and benefit cuts" were by far the most common reasons for people seeking help. People were very reluctant to claim UC, even when in dire need, because of the delays.

'Most people are reluctant to apply for benefits especially now big bad Universal Credit is around so if they have any money or if they have any way of avoiding it, while they are hoping to get another job, so many people will do anything. A man losing his job is quite happy to take a job at the fish farm or a job in one of the hotels. It's anything to keep going whilst they are looking for something.' (Gatekeeper)

'I'm going to be doing a home visit tomorrow to somebody who can't afford to come in, they just don't have the money at all, they've got absolutely nothing at the moment – they've just started on Universal Credit and they've got, I think, £160 to live on for the month because it's a monthly benefit and that's going out on bills, obviously. So, I'm going to have to take food parcels and shopping to him as well. Partly [it's because of UC delays and partly because] he's been in hospital as well, he's had mental health issues, there's a lot of debt involved. [He's] probably had an advance payment and he's got housing arrears and he's got council tax arrears; I don't know how many deductions. So, he only got it two days ago and he's had to spend it all, just on bills, and he won't have any food and he's facing a potential eviction and it's just a nightmare.' (Gatekeeper)

This last quote is a reminder of the difficulties faced by those with mental health conditions, which came through so strongly in the Perthshire case study interviews.

Chapter 3 outlined some of the reasons for lower claimant levels in rural Perthshire. A similar pattern was found in Harris. Elderly people were particularly resistant, and many interviewees highlighted the importance of carers and others in providing support and raising awareness. Stigma, visibility, privacy and pride were mentioned repeatedly in the interviews, particularly in relation to the older generation.

'I just think that they've still got that notion in Harris of having that strong cultural identity of people just managing and just getting on with stuff.' (Gatekeeper)

'What I've found in particular, is with the elderly, a reluctance to claim what they're entitled to and in a recent case, where this woman was absolutely adamant that she wasn't wanting basically to apply and all through their lives up here, they've managed on what they had and they'll be fine. Before, when Pension Credit came out in particular, we had an awful job getting the elderly to claim Pension Credit and it opens up, particularly the guarantee portion of it, it opens up so many more doors for them. But it's just this attitude of, we have enough, we're fine, we've managed.' (Gatekeeper)

'The carers are in for such a short time nowadays that they've not really got the same opportunities to discover what's going on. They [older people] pretend everything's okay for 15 minutes while the carer is in and then go away. And it's just that part of the Hebrides psyche that's too proud to ask for help, too frightened of what might happen if you ask for help, get taken into a care home or whatever. So, there's that element of hidden poverty.' (Gatekeeper)

Many respondents believe that younger generations may be less reluctant to claim:

'I find attitudes have [changed]. Nobody feels stigmatised anymore because everybody seems to be either in low-paid jobs or on benefits or retirement pension or whatever and they are coping. They are managing but aren't afraid to say they are struggling. If they do say and if somebody can suggest a way of easing their burden then that happens and these doors are open to them. In the past you had to really prise it out of people if they were struggling with anything.' (Gatekeeper)

The elderly were said to have a distinct relationship with savings and debt, compared with the rest of the population: due to fear, they were less likely to have debts, and they were more likely to have some savings. Having a minimum level of savings, to cover their funeral costs, was very important to many:

'They want to make sure they're not a burden on anybody and that starts from having that £5,000. They've all worked it out between themselves. And that'll be then, "I'm making sure that I don't touch that £5,000 because that's what's going to bury me because I'm not going to be a burden on anybody".' (Individual)

On occasion, people in need might be willing to talk to trusted contacts, voluntary and community organisations or to advisers who are not from the island:

'I think a lot of my referrals came through health visitors which has been so helpful because again talking about the stigma and things, you already trust your health visitor because you've known them for so long or you trust your community nurse that comes out to you every week. I've definitely felt like I've got to more people there who didn't have anything, compared to referrals coming from other places or people who would be more reluctant. So that's definitely helped and then the likes of TIG [Tighean Innse Gall][4] definitely does make a big difference – it is really good.' (Gatekeeper)

'There are definitely a few clients who recognise my [mainland] accent and stuff, who are not from here, who have asked for me again and saying, you've not been here that long, can I see *you* if I come back because I don't want to see someone who knows so and so. That's maybe one of the reasons that Harris come to Lewis [for more confidential advice].' (Gatekeeper)

There was administrative evidence too of lower rates of uptake of crisis grants, community care grants, council tax benefit and council tax reduction in Harris, for reasons unknown:

'Harris has our lowest take-up of for crisis grants. Now I'm not sure necessarily that doesn't say there aren't people in crisis, it just means that people aren't utilising that route quite so often. So, from that side of things, we don't see many community care grants either.' (Gatekeeper)

Visibility and stigma are concomitants of many more positive aspects of life in smaller communities, as discussed later in this chapter.

One further issue which emerged was of households that are financially vulnerable, and perhaps even facing financial hardship, but which are ineligible for state support.

'People in low-income families are always the ones brushed off in society, I think. As I said, people who are on benefits, who have been on them long term and are used to being on benefits, they budget; and people who are high earners don't have these financial worries anyway. But low-income families really are getting overlooked. Yes, it can hit them the most. They are maybe earning a little bit but just too much to be able to get any help with the rent or the council tax so they struggle. Then if they have a bill like the washing machine or anything like that then they really struggle.' (Gatekeeper)

Welfare conditionality has increased in recent years, excluding more people from eligibility for benefits, alongside policies of not increasing benefits and thresholds in line with inflation. As a result, more families and young people face financial risks without the help of social security, as the previous quote suggests. Many more experience financial vulnerability, for example, to unexpected bills. Another result of these thresholds is the poverty trap creating disincentives to work.

Staff at the Stornoway Job Centre were said to be helpful and flexible, as were the local council and social landlord. Being a small community where people are known does tend to help:

> 'generally, agencies take an approach which is sympathetic towards folk and understands their circumstances and they're our neighbours as well as having a common debt problem. I have to say that's a genuine approach. The only thing that I can imagine about poverty, you probably live in fear most of the time, what's the next letter, so I would say we genuinely try not to add to that and we need to think about that in the poverty strategy, about the culture that we have and to maintain that.' (Gatekeeper)

A lot of partnership work is being carried out to share information and coordinate responses. The Community Planning Partnership (CPP) seeks to coordinate much of this effort, within the framework of the Outer Hebrides Anti-Poverty Strategy. The CPP is also very active in advocating rural proofing, to ensure that national, regional and local policies are tailored as far as possible to have maximum impacts at the local level.

Support from voluntary and community organisations

Support from voluntary and community organisations was mentioned less frequently by respondents in Harris than in the other study areas. Nevertheless, this type of support was highly valued by those who received help: the importance of advice agencies has already been highlighted.

> 'You will see this, and I know [X] is an example where somebody has turned their life around because some of the debt has been sorted out. One lady came to me and said I wouldn't be here; I would have taken my own life had it not been for the advice that your colleague gave.' (Gatekeeper)

A recently opened foodbank in Stornoway has tried to find ways of reducing the visibility of people using their service:

> 'the Fareshare [foodbank], they're put into shopping bags, normal shopping bags, and say two or three of these will be in the van of the, maybe the social worker or the health visitor, the child support team

and if they see somebody they think that they need it, they'll just give it over without a lot of paperwork being done.' (Gatekeeper)

To try and reach more people and address the issue of visibility, the CAB has been added to an app used by the National Health Service which enables them to make referrals to other relevant services. We were told people in debt tend to bury their head in the sand, exacerbating the problem. Support agencies reported using ring codes, or arranging to ring at a certain time, in order to ensure that clients answered the phone.

Community land trusts were seen as central to the positive turnaround experienced in Harris and were seen as a big part of the solution in the future. Respondents stated that the community buyouts had created a greater sense of community confidence, as well as more concrete benefits such as housing. People were wary, however, of expecting the trusts to do everything, as there was recognition that it wasn't in their power to do so, and they often lacked the necessary resources. The local authority and other public agencies were still expected to fulfil their respective duties.

'The community trusts, they can identify locally, really locally, what is happening and what the problems are, what are the issues. So, that has been a big driver. Now, it's a pity that they're not financed better

Behind closed doors: foodbanks in the study areas were often hidden away in discreet locations

because, had they been financed better, I think you would see a lot of the issues we're talking about in terms of what do we provide for childcare, what do we provide for disenfranchised families. You would see them targeting those areas because it would become apparent to them this is an issue for us. People would come to them and say, "What are you doing about …?" I think the community trusts have a massive role to play but the problem is getting, probably finances … for me, they are a natural route to development, to economic development in these communities.' (Gatekeeper)

'on the one hand the community council is the official democratic body within a community that local authorities have to deal with but on the other you have this alternative power centre which is the community landowner which, in practice, wields an awful lot more power.' (Focus group)

Many of the land trusts have an independent source of income (primarily from renewable energy), but many others don't, and this limits their capacity to take forward initiatives to address local needs. During the pandemic in 2020–21, those with their own income source were able to initiate local measures to support their community, whereas others that were reliant on tourism income furloughed their staff. This is discussed further in Chapters 7 and 8.

In the Bays area of Harris, where residents are currently considering community ownership, many in the community were hopeful this would happen in the near future: it was seen as the way to secure development on the eastern side of the island. It was believed that community ownership would bring new life and vibrancy to the area, especially as the current estate business was understood to have sufficient income to make a new ownership model financially sustainable.

'Community land issues might be really vital here in amongst that, it's that kind of timing isn't it? Harris' plan and that folk feeling empowered by the things going on, and the branding maybe improving.' (Gatekeeper)

Taking on greater responsibility offers an opportunity for voluntary and community organisations to diversify their income sources (if properly funded to get established) and to respond to local needs, but there is also the risk of these organisations losing their independence and mission focus. This raises the question of how appropriate it is for the responsibilities of the local authority to be transferred to voluntary and community organisations and on what financial basis?

Apart from a concern about the reliance of voluntary and community organisations on older volunteers, these organisations also face growing funding pressures at a time of rising demand and limited capacity, as noted in Chapter 3. However, in Harris there may also be signs of the emergence of a more collaborative and mutually beneficial partnership approach:

'We have that Harris Forum now; we've set the Harris Forum up and that is really working really well but that's come from a very strong base where you've got two community trusts that are bringing in initiatives all the time. You can even use them in areas of focus and say, well, statutory agencies are saying, right, we want to target these people in the community who are really struggling for this, right, we'll give you X amount and you focus on it – you know who they are, go and deal with it.' (Gatekeeper)

In fact, the interviews in Harris revealed many examples of successful and effective joint working, cross-referrals and informal partnership arrangements, involving health practitioners, CAB, TIG, electricity providers, Home Energy Scotland, Macmillan and, of course, the local council and the CPP. A highly regarded example was TIG's close working with a host of other organisations on the back of energy advice and home insulation:

'I guess from my part it's a holistic approach that we try and take and the referrals are so widely important to other agencies, we feel because for income maximisation you have got the financial inclusion section in the council and WICAS [Western Isles Citizens Advice Service] and then, when you have got other concerns, if someone is diagnosed with cancer you go through the Macmillan route. There are a whole host of things, if you come across someone with mental health issues, these types of difficult situations, we can make the referral into other agencies locally, drug and alcohol abuse and social interaction with people. When you enter a property [to assess energy use] you don't know what you will find but you always come back with lots of things you can do, particularly from the referral mechanism.' (Gatekeeper)

For many, the strategy they pursue in times of financial hardship is to focus on ways to maintain independence and self-reliance. Only when this strategy failed would they seek help from one of the organisations mentioned earlier, such as the CAB.

From the perspective of the many groups and organisations who seek to offer help and support, there remains the considerable challenge of how to reach the 'hard to reach' and how to ensure help and support is available to those in scattered rural communities at some distance from service centres. Many providers are pursuing digital approaches to reaching rural residents, and some are continuing to offer non-digital contact by telephone or through physical outreach activities, such as drop-in consultations. This plurality is important because some people do not have digital access or capability, but, like in Perthshire, digital by default approaches are becoming common nevertheless.

> 'Well, they could phone in and make an application over the phone … and we would make the payment direct into their bank, so if somebody was unable to come in and pick up cash then that's fine, because we have quite a lot of that through the islands into Uist, for example, South Uist, it's very hard for them to get into the offices quickly.' (Gatekeeper)

The general view of respondents was that, with a few exceptions, levels of severe debt were low, and much more common were low levels of temporary debt. This was the experience of the local CAB office, but respondents also acknowledged that debt would not easily be spoken about locally because of a sense of privacy and a desire to be able to 'manage'. Should someone have high levels of debt, they were more likely to seek help outwith Harris, either going to the CAB in Stornoway or even seeking help off-island. This view was supported by the respective CABs.

> 'Very few people come to this bureau if they have got financial problems. We are such a small community although people are more likely now to come with small debts like maybe catalogue debts or a little bit of council tax debt. Anything that they deem in their minds as big they would go to the big anonymous bureau in Stornoway. We have been open for 30 odd years, and that hasn't changed. The only thing that has changed is that people will come if they have a few hundred pounds of debt and they need help to manage repayments, anything in the thousands, I would say, wouldn't come.' (Gatekeeper)

The credit union in Stornoway was heavily used by older people on a pension, who wanted the ability to save a small amount on a regular basis towards routine costs, such as visiting family on the mainland, or Christmas. This helped them mitigate against financial vulnerability.

The various agencies on the island were said to be very helpful and understanding, and again there was evidence of a lot of joined-up work to try and solve debt where it existed. This was seen as one of the positives of being an island community, where people knew each other and the agencies were small enough, close enough and flexible enough to be more joined up in their approach.

> 'Generally, and this is part of the advantage of having a local, not being in Glasgow, not being in a big city. Generally, people are pretty good in terms of HHP, the agencies are pretty good, the council are very good in terms of working with people to try and help them sort out their problems. Because it's so local, it's maybe someone's cousin, it's in the same finance department or whatever, they work, you tend to find people are very, very good.' (Gatekeeper)

Support from friends and family

Informal support from family, neighbours and friends is particularly important in Harris, as found in earlier studies of the Western Isles (Shucksmith et al, 1996), and indeed this was a recurring theme in our interviews. It was, however, very much tied up with the sense of place, as the quotes to come illustrate: the mutual reliance of neighbours around crofting, traditions of visiting and the awareness that comes with living in a small community with strong historical family ties. The importance of place is therefore discussed in more detail towards the end of this chapter.

> 'It is good living here. I wouldn't move away. I wouldn't move away. I can't put into words why not. I mean, I like to get away to the mainland but then after a couple of days, I'm like, I want back. Yes, it is the community spirit here. With us, everybody would be round you. And phoning and asking and, "Are you alright? Is there anything I can do? Do you need a lift to the hospital?" So, everybody's offering everybody help and lifts. Whereas Inverness, mainland, nobody knows you. You're just invisible. I just feel that up here, it's just, the community spirit is nice. Some people can say, "Nosey, nosey, nosey". But you ignore that element of it because when the chips are down, people do help.' (Individual)

> 'I am not sure how widespread it is but it certainly seems to be that everybody has got some kind of access if not personally in their own homes then a neighbour, a friend or a family member will help them.' (Gatekeeper)

> 'I think the volunteering rate is really high. A lot of people don't consider that they're even doing it but at least they are, just helping out neighbours and things like that, that's more community based.' (Gatekeeper)

Alongside this strong and ubiquitous narrative of informal support from family, friends and neighbours, there were also some suggestions that this source of support was dwindling as a result of different patterns of living and wider changes in society.

'Visiting, that's declined a lot. A lot of that has stopped. I don't know why it's stopped but people don't have big families living together anymore like you had Granny and Grandad, Mum and Dad, and then us lot within the house as well, people were always in and out of each other's houses. I think the telly, the internet, people don't go out to the shops the way they used to. You would always be at the village and that would take you half a day before you got speaking to everybody.' (Individual)

Housing, cost of living and access to services

Housing was the issue that everyone cited as the key difficulty that the island faces at the moment, and the key priority for the future, as discussed previously. In 2018 there were 1,169 dwellings in Harris, with a significant proportion being vacant (10 per cent compared with 8 per cent and 3 per cent for the Western Isles and Scotland respectively), or used as second homes (12 per cent compared with 5.6 per cent and 1 per cent for the Western Isles and Scotland respectively). In the south of Harris, the level of second home ownership was 15 per cent of the total stock (Comhairle nan Eilean Siar, 2019). Recent research carried out by the West Harris Trust found that 21 out of 38 properties either recently built or in the process of being built were holiday homes or self-catering properties (Russell, 2020). The same research also found that a two-bed property put up for sale around the time of our research was on the market for offers over £385,000, placing it outside the affordability range of most local residents who may be in need of housing. There is also generally a low turnover of housing stock. In 2018 the Outer Hebrides had a decrease in the volume of house sales of 19 per cent, compared with 2.3 per cent nationally.

Those most likely to struggle to purchase a home were newly formed young households: single-person households, couples or new families. In recent years attempts have been made to bring small housing units into the market, but there was still a real shortage:

'Accommodation is a real issue, housing is a real issue across the islands, but in Harris … for someone who needs a one-bedroomed flat or something smaller for a small family or something, they're really, really going to be struggling.' (Gatekeeper)

'I don't know, the lack of housing seems a very big thing. I live with my parents. There's no hope, there's no houses at the moment.' (Individual)

Harris has a dispersed settlement pattern, with crofts in rugged terrain

The focus groups confirmed that young people struggle to get housing, which means that the islands effectively "export poverty" to the mainland, where it is easier to secure affordable accommodation on a lower income. We spoke to young incomers to the island who had experienced difficulties with finding somewhere to live, and only succeeded because of "luck and timing". Respondents questioned why more social housing wasn't being built to relieve the pressure, or why more wasn't done to bring the high levels of empty properties back into use. An alternative solution mooted was to actively develop holiday lodges/pods which could be used by tourists instead of domestic properties. Although it was acknowledged that tourism impacted negatively on the supply of affordable housing, the solution was not seen as restricting the levels of tourism but increasing the supply of housing. It was also noted that it is locals themselves who are selling or have sold houses/plots of land for holiday homes.

It was suggested that "out of the box" thinking was needed to meet the demand, for instance, using housing pods, welfare units or even one of the accommodation ships that are used in Shetland to house oil workers. These were not put forward as permanent solutions but as interim measures while a long-term solution is developed. In some of the crofting townships the level of second/holiday homes was reported to be very high, and respondents would sit and count the houses in their community that were not permanently occupied:

'If you look at some of the villages, you count the number of houses which are lived in or Airbnb, I think in Northton it's about 12 houses are permanently lived in, 12 or 14, and the rest are rented out. It's a massive, massive problem. Probably the biggest issue at the minute is housing. I mean if you want a job, anyone can work here. There is work. Where do you live?' (Gatekeeper)

Some of the holiday properties double up as winter lets for locals, although there was general agreement that the tourist season is getting longer and longer, resulting in fewer long-term winter lets. One respondent, however, raised concerns about how many holiday homes would continue to also be long-term winter lets in future due to a change in the legislation and the introduction of the private residential tenancy which will make it harder to have fixed short-term tenancies over the winter period.

According to respondents, there are two general categories of second/ holiday homeowner. The first are family members who have inherited the property from the original occupant, who were said to want to keep the property so as to maintain their connection to the island. The second category are people from the mainland who had bought the property on the open market, either for their own use and/or for renting out. These owners, in particular, were seen as the reason for house prices having increased beyond the reach of young local people (and others) wanting to get into the housing market.

'People come up. Biggest thing I've seen, people come up, love the place in summer, buy a house and prices are through the roof. Cheapest house we've seen advertised was £85,000. It's just up the road here. And it needed a new roof, a new floor. You wouldn't have got a mortgage on it from the start. And it sold for £120,000.' (Individual)

There were other aspects to affordability that were raised by respondents, from getting access to land to construction costs. Sites were said to be expensive because of the demand for housing from the mainland and for tourism. On a sparsely populated island this is perhaps contrary to expectations, but much of the land would be very challenging and expensive to build on. As building and materials technology advances, it should be possible to develop more innovative building methods that would be suitable for the Harris terrain, bringing the cost of site preparation and construction down, and opening up sites to affordable development.

Much of the land in Harris is now owned by the community. However, ownership is only one factor in accessibility, as often the power over land use rests not with the owner but with crofters, and the Grazings' Committees in particular. This was identified by several respondents as a constraint for housing development, particularly as the common grazings is often some of the more accessible land in construction terms. The community land trusts have managed to facilitate the construction of new social housing, and they were applauded for this, but it was felt that it still wasn't enough to meet the need.

> 'You don't see a lot of house moves or changing tenancy because there are no tenancies to give. So, if you get an HHP house, which is our main housing association, you keep it, you're not going to move out in any hurry. So, they aren't moving quite so much once they get these tenancies.' (Gatekeeper)

But there is a shortage of social housing too, in some cases to such an extent that people do not bother to put their names on the waiting list. And when a new affordable house becomes available, many people were said to be applying for it. There is little turnover in stock and high demand.

> 'One of the flats came up. I think there were 20 or 30 applicants for one flat. So, there is a demand for it … the young people aren't getting the opportunity to start experiencing what housing is and having housing and the discipline of paying rent and all that or paying a mortgage. That doesn't happen because there is nowhere for them to do it. So again, they're staying with parents or they move off the island for work if they want to work.' (Gatekeeper)

> 'a lot of people here don't bother going on the housing list because, "Why am I going to put down for a single-person flat in Tarbert because a) there aren't any, b) there's no plans in the next five to ten years to build any so what's the point?".' (Gatekeeper)

Housing quality was also highlighted as an issue, with the impact this had on quality of life and heating costs, for instance, also being outlined, although respondents acknowledged the significant improvement and insulation work that had taken place in recent years. TIG was once again recognised in particular in this respect.

For older people, issues of fuel poverty were said to be compounded by the lack of housing options, having to stay in overlarge properties due to a lack of alternatives. Other issues mentioned were lack of mains gas, heating methods and housing design and materials which made a property on the one hand cold, and on the other hand difficult to insulate. The exposed nature of the islands and more extreme weather conditions added to the difficulties, and made properties more costly to heat and maintain.

> 'It's costing us a fortune the house we're in just now. It's all electric. [The council] thought it would be a great idea, of this new-fangled infra-red

heating. When we had it on, we were using nearly £7 a day, simply to just to keep the house … well, it wasn't even warm.' (Individual)

'heating this big huge house has been a problem, even though they're really lucky, they've got their house, they don't pay their housing costs and stuff but practically they can't afford to heat their whole house or it's actually not accessible now with their needs.' (Gatekeeper)

'There's fuel poverty, there's the standard of housing, unless it's social housing in Harris … you will have three-foot walls, stone buildings, leaking energy as you sit there and you can't afford to maintain them.' (Gatekeeper)

'Of course, the way we are, out in the islands here, our houses here, we've got a loch on the one side, the sea on the other, so we get the blast from the wind from both sides.' (Individual)

Debt as a result of fuel bills was also raised:

'I think we're about the second when it comes to fuel debt in Scotland but when it comes to severe fuel debt, I think we're first and when I say first, it's the worst, not best.' (Gatekeeper)

In recognition of the costs of heating homes, especially with infra-red heating systems, the CAB has applied for funding for fuel vouchers which they can give out to households struggling to meet their heating costs.

As in Perthshire, respondents noted the higher cost of living associated with living in a rural area.

'I think the cost of living has undoubtedly got … it's more and more expensive to live and we just have not kept the pace of that recently. So, people have much less disposable income, so we're seeing a lot more individuals juggling debt about, quite often amongst high-cost lenders.' (Gatekeeper)

'Shopping, food is more expensive – we don't have Aldi or Lidl. No, and this brings up this supplementary thing that we don't have choice, so you go to that garage or there's not another garage. You don't get the Tesco free petrol stuff. Well, if you go to Tesco and you get the voucher, I get a voucher every single time I go to Tesco, I get a voucher for my money off fuel but we don't have a Tesco garage. From Argos, they won't deliver certain things and you don't get the benefits that maybe a city centre or a town centre and postage costs, delivery costs.

It's not just pence more, sometimes it's twice as much as the cost of the actual thing.' (Gatekeeper)

There were a number of different aspects to higher food costs, with a lack of any of the cheaper supermarkets (Aldi, Lidl) on the islands causing frustration for respondents. In other areas, these shops are well used by people, particularly those on a tight budget looking for ways to make ends meet. Food costs are also higher, either because of the transport required to access a supermarket (minimum 70-mile round trip), or because of a reliance on the local shop, where one exists. For the elderly, the group most likely to

Service provision is a challenge for the dispersed population: Valtos post office

be in poverty, there was a greater reliance on the more expensive local shop because of the distance to the supermarkets in Stornoway. They also don't have the same option to buy in bulk to keep costs down. This sometimes leads to difficult decisions:

> 'We've got the Co-op and we've got Tesco's and the local shops are really expensive because they don't have the ability to bulk buy at Tesco's and elderly people in Harris mostly shop in the local shops. So, they're buying less food and they're having to make decisions about heating the home and having food which a lot of pensioners unfortunately and regrettably are having to make.' (Gatekeeper)

People did complain about lack of choice, not just in relation to supermarkets, but for all sorts of goods such as electrical items and clothing. Online shopping has therefore increased, putting further financial pressure on local shops.

> 'I think probably the other thing is, if you're on the mainland … you certainly get more of a choice, more shops to go around and if you're looking for something cheaper you just go round the shops there.' (Individual)

Lack of choice locally means people rely on goods bought off-island, particularly if it is a specialist item. Delivery costs were said to be high, even on essential items such as spare parts for boilers and other household appliances, as well as the appliances themselves:

> 'If you have to have things transported here, you get big transport costs. Some companies charge a lot of money to send things up here. You know I was having fights with people on the phone about, you know, if it's a small item why is it costing –?' (Individual)

Because of poorer access to goods on the islands, there is a greater reliance on catalogues, and several people cited excessive use of catalogues as a cause of debt. Catalogue shopping has persisted more in the remote areas of Scotland, not only because of lack of choice locally, but because catalogues often offer free postage to the islands if you buy on credit.

> 'We had a client come in yesterday … and it was just high-cost lender after high-cost lender, catalogues or something – you probably don't see so much on the mainland but it's huge here, because they'll do free delivery on credit. So, people like that because you can't just nip into a shop for stuff here.' (Gatekeeper)

There was felt to be increasing peer pressure, especially on school-age children, to have the right clothes, shoes and phones, leading to increased living costs, sometimes beyond the affordable level. School was also said to bring with it additional costs, such as school uniforms and now IT and internet for schoolwork at home, especially during the COVID-19 pandemic. While these are costs that people in all areas face, the lack of local choice and consequent higher prices, and possibly delivery costs on top, make the overall cost higher. Although free school meals and clothing vouchers are available, they were said to not cover all the additional costs people incur, and there is stigma associated with these.

Travel was seen as an essential feature of life on Harris, but there is limited public transport, and so a greater reliance on private car ownership. Travel costs were high – from fuel costs to the greater distances travelled, and from tyre costs to car servicing (which may require a mainland trip):

> 'The cost of fuel for your car, the cost of tyres, the cost of everything here. You have to go to the mainland to get your car serviced and that's a ferry trip which is no longer the road equivalent tariff. ... And usually I go to Inverness to get my car serviced or if there's any of these dreadful problems that come up on the computer nowadays and they cannot be fixed unless they are plugged in for two minutes in at Arnold Clark in Inverness. So, that's an extra few pounds, an overnight trip. So, that's another hotel ...' (Gatekeeper)

As well as incurring these costs for daily life (work, shopping), there was the added cost of going to the mainland for things such as holidays, family visits and medical appointments. The introduction of the Road Equivalent Tariff (RET)[5] has made a big difference to local people, making travel to the mainland and back more affordable. But there was a sense that while RET works well for locals, it is also subsidising wealthy visitors to the islands. The net effect of this was to fill up the ferries in the summer at the exclusion of locals needing to travel.

Public transport is limited, and respondents recognised the difficulty in providing public transport – it was a necessity for those people that relied on it, but their needs are varied and so having a service that could cater for everyone was very difficult (and costly) to achieve.

> 'Transport is always an issue ... Basically, if you're living in a rural house you need a car, that's a part of life, and you can't expect public transport to be geared to everybody's needs. I mean very often these buses are going up to Stornoway, very few, sometimes empty.' (Gatekeeper)

Children in particular were seen as being disadvantaged because of the costs of travelling to the mainland to engage in sporting and music events and competitions, engagement in which is taken for granted on the mainland. Most of the leisure activities that people engaged in on Harris were centred around committees, community events and visiting, as there is not the same availability of more organised and costly leisure options. This has the added benefit of providing activities and services through community and social enterprises.

Mobile and broadband coverage are not comprehensive across Harris. Some areas were reliant on satellite broadband, and others had broadband provided by fibre. As well as issues of variable provision, there were also comments made about the cost of the services available. Limited broadband, and poor mobile coverage, were felt to add to social isolation in some cases. This will have been heightened during the COVID-19 pandemic, which has seen unprecedented use of social media and video calling as a way of maintaining contact with neighbours, friends and family, and also medical and other services.

> 'Yes, I've got the internet but I haven't got a proper mobile signal. I have to go upstairs. There's a mast over in Scalpay there. I have to go upstairs and stand in the window. If it's a good day I'll get a signal with my mobile.' (Individual)

> 'People have to work from home and there are some areas that still don't have the fibre, the fast fibre. Somebody who moved back to Harris and was going to build on the west side, who has got a croft on the west side … but her job depends on the internet, fast, fast internet access, and it's still not available in that particular area.' (Gatekeeper)

> 'I suppose if you're into having broadband … you have to pay a lot of money to have satellite and then you pay for your landline as well, so it all becomes quite expensive that side of things.' (Individual)

The focus groups confirmed the limiting aspect of broadband availability and quality, with one discussion noting that some businesses had had to close in the Bays area due to poor connectivity. Connectivity was seen as "really holding the place back". It was also said to be much more vulnerable to extreme weather events which can be fairly regular:

> 'it feels like any time we have a thunder storm or like lightning the other day, we completely just go off the face of the earth … those kind of things do impact. We only just got our phones back up and running

yesterday and that thunder storm was last week or something. … That kind of connectivity is so important for [our business].' (Focus group)

The increasing move to provide a range of services online, such as banking, healthcare, benefits and advice, will mean that those with limited broadband will be at a disadvantage. There are public internet access points, but these are not always suitable, requiring people to still travel. The impact will be most felt by the elderly, school children, people on benefits and people without their own transport.

Many services were being increasingly centralised, with the loss of small schools, banks and local shops. Access to some services was becoming difficult as a result. Access to banking and post office services in particular was more challenging and the effects of centralisation of these services was only beginning to be felt. Among older respondents, access to medical services, especially in times of an emergency, was of concern.

Although the level and frequency of some services had clearly declined in recent years, other things had improved. Infrastructure improvements were spoken about as having contributed to change in the islands, usually for the better. There had been improvements to the electricity supply, for instance, which meant a more reliable service that was less vulnerable to outages. The dualling of the road with European funds over the last 20–30 years was said to have transformed travel on the island, cutting travel times significantly. The building of the Scalpay Bridge, by contrast, was felt to have contributed to the decline of Scalpay, as it had enabled people to shift their focus from Scalpay to Harris itself for work, shopping and schooling.

'building the bridge seemed to … it's been the opposite, I think there's been a bit of decline there partly to do with the bridge. Then that brings new issues for people that are struggling because they've then got to drive to their doctor or get a bus … Or the shop, because the shop has closed in Scalpay now. That really changes things for elderly people.' (Gatekeeper)

Place

'They say about history, history of a place is a history of all the people in it.' (Individual)

History and connection to place were common themes raised by all respondents. Reflecting on the changes in Harris in recent years, people noted not only what has been gained, but also what has been lost. Linked to the importance of family and friends in times of difficulty, a sense of community was stated as a very important feature of Harris life, and while this was felt to have declined in recent years, it was still very much a feature of island life. This is consistent with findings from the work carried out by the CPP using the Place Standard methodology.[6] Scores in Harris were highest for sense of community, identity and feeling safe, and in each case they were higher than the average score for the Outer Hebrides as a whole (Outer Hebrides CPP, 2017).

The influx of incomers to Harris was portrayed as making the community more diverse, for good and ill. It was acknowledged that incomers, once seen in a very negative light, have been the thing that has "saved" the island, without whom Harris wouldn't have undergone its recent revival:

> 'I mean, one reason is that the population has dropped. You know, despite all the developments, despite all the effort, the population has continued to fall – to decline. If it wasn't for the incoming population we would be really in dire straits, because we would have an ageing population and very few employable population.' (Gatekeeper)

Crofting is a key element of the identity of Harris, and has defined its culture, landscape and settlement pattern. Everybody talked about the crofts in their community, who was still crofting and what aspects of crofting they still engaged in:

> '[Crofting] has changed. That has changed enormously. There's very few now who are full-time crofters. There's a few who are full-time crofters, but the number of cattle, for example, has just dropped. Nobody milks a cow nowadays. Yes, and again, because of the ageing population, some crofters have stopped, you know, because of the age of the crofters are now getting older and older, and they are not able to go into the hill and it's very difficult to get other people to pay to go to the hill for them. That's another issue they are having … they still put them [the sheep] out to the hills – the ones who have them. Trying to get people to gather them in is really a problem because there's not enough young people, so they have to run hills and they are too old to do it anymore.' (Gatekeeper)

The sense of connection to place was very strong among interviewees. Those who had left the islands when they were younger spoke of being drawn back there to live.

> 'You'll get this from your talks with folks in Harris, but I get the sense of folk that a lot of community pride, that's why you keep a second home in Harris, the identity with that island and you wouldn't let that house go.' (Gatekeeper)

Sense of community and a slower pace of life were part of this draw, but it was also something more than that, something less tangible. Going away was seen as important, so long as there was a way back for those that wanted to, to come back. And because people felt the pull of the islands themselves, they were able to understand why people from the mainland also felt it.

'Because you do see a lot of people move away and they come back. Something that draws you back. People come back when they're older.' (Individual)

'What made me come back? I think it was just that sort of … It was always a pull for me to come back here, I think. Even when I was a student, I thought oh, I would love to go and open a gallery before there were any galleries at all thought of in Harris.' (Individual)

'If you're from the island that's all you want to get back to it at some stage in your life, that's the majority of people so they tend to be in their old age after retiring, if you're from here.' (Gatekeeper)

'People do come back … we were really happy living in Glasgow but then when we had kids, we realised we wanted our kids to have a better quality of life where they could run the croft and run around, riding their bikes safely.' (Gatekeeper)

'They're able to take use of the local environment, they want to live in a remote rural island location and it's absolutely stunning and beautiful, so why wouldn't you?' (Gatekeeper)

Respondents still very much valued and recognised the close and supportive community they lived in: this was seen as something special about the islands, and contrasted markedly to the anonymity of the mainland. This could be seen as a drawback – everybody was said to know everyone else's business – but on the whole it was positive, making it a more caring and helpful community. People spoke of being able to safely leave cars and houses unlocked, low levels of crime, unprompted acts of kindness in times of difficulty, among other benefits.

'The plusses, I'd say the quietness, the security of not having to lock your car and not having to worry about your door being locked all the time. Lack of crime. Just all of that is a big plus. But … You know, it is changing and with the more people that come in you don't know who the tourists are even. I think in the summertime people are a bit more security conscious … I think looking back to how it was, it was

Fisherman on the east coast of Harris

a tremendous cohesion with the community spirit and people working together and looking after each other ... somebody was talking about ... passing places and widening the roads, and somebody said we don't want the roads widened, these passing places are places, you know, where we stop and talk to our neighbours.' (Individual)

There has been a change in the community – people know their neighbours less than they did previously, people have moved on or died, new people have moved in and there isn't the same interaction as before. There was also said to be less *need* to have the same level of cohesion – in the past people needed to work together to tend to the animals, to cut peats or to bring furniture up from the boat to the house. Fewer people are crofting, furniture is delivered in a van and more people have cars and are able to be more independent.

'In these villages everybody helped everybody. Yeah, if you had a cow you'd give milk round the village and free eggs and all that. But now there is no community because all the older folks have either died or the families have moved away, and different people have moved in.' (Individual)

People knowing your business was usually seen as a very positive aspect of the community, because it meant that people looked out for each other, and took care of each other in a way that they didn't think would happen on the mainland. It meant that it was easier to care for those less fortunate, and a place was found within the community for people with additional needs that in a bigger community was seen as not being possible.

> 'It's not so hectic, and some people, it can be a good thing and a bad thing, everybody knowing your business. The community spirit here is good … my [relative] stayed up in Tarbert and he had severe dementia. And everybody watched out for him.' (Individual)

We have already noted that the tradition of visiting was said to have declined, and some reasons were suggested for this. Smaller family sizes and fewer multigeneration households were also seen as factors – the more people who lived in a property, the more reasons there were to be visited.

> '[the visiting tradition is] not what it was at all … it's just that people don't need each other anymore. They used to work together, if you were crofting you were working with others who had sheep and you also needed people to help you with things. People were closer together whereas now, well they still need help from others but not to the same extent.' (Individual)

Linked to the decline in the tradition of visiting was a reported increase in the sense of social isolation, which was also exacerbated by declining population numbers and the loss of people from the island. People were less likely to know their neighbours than before, assuming that neighbouring houses were actually occupied year-round. Those most at risk of social isolation were said to be older people and young people (children and young adults up to their early 20s). Increasing centralisation of services, leading to less local delivery, was thought to make matters worse for older age groups, and conversely better for children at least, with schools being amalgamated. One key way that some people avoided social isolation themselves was to get involved in volunteering in one way or another – being on committees, volunteering for Crossroads (to provide respite care in the home to carers of the ill, elderly or people with disabilities), running the local football group and so on. This was a strategy used by young and old alike.

> 'For me, it's allowed me to get involved in community groups, youth clubs and football. … You really have to get involved here or else things aren't meant to be done.' (Individual)

Everybody we spoke to mentioned how much busier the island was because of the number of tourists visiting and because the season had significantly extended. It was said by some to be almost year-round. This is very important economically for Harris, but it has its downsides. The impact on housing has been noted previously, but there are other impacts:

'I mean, it is, in the summer here, it is hectic sometimes. But like I say, the ferry, you can't get booked on the ferry … I had to put the car away the day before, there was space on the ferry the day before and one of the crew took it off the ferry in Uig and then I had to go on as a foot passenger. So, if you had a family emergency, you'd have to go as a foot passenger if you couldn't get the car on.' (Individual)

The roads were also busier, causing significant delays in the summer. Much of the infrastructure, albeit drastically improved over the last 20 years, was said not to be able to cope with the volume of traffic and visitors in the summer:

'How busy it is in the summer. Busier, busier every year. … It's just getting worse and worse. People don't really … I don't know. Do they not realise that we've got stuff to do?' (Individual)

'It's a victim of its own success really. The infrastructure is really struggling, not only housing but for roads, the road network is a nightmare. Yes, just the amount of volume on the roads. So, if you're trying to get anywhere in the summer, you have to add 30 or 40 per cent on to your time to get anywhere.' (Gatekeeper)

Attempts are being made to extend the season further and to even out the visitors over the year to reduce the seasonal variation and impact. This would give more secure income and employment year-round.

One of the most frequently mentioned, and important, issues for Harris residents is how to provide help and support for older people requiring social care. While many looked back to past times when family, friends and neighbours would have provided such care informally, there is an increasing reliance on formal care provided by public, private or voluntary sectors, whether in care homes or in the community. Care services have faced funding challenges under austerity to such an extent that carers now are not given adequate time to care properly, in their opinion, for each person they visit, a problem exacerbated by long travel times.

Local services, and in particular services delivered at home, were seen as crucial for reducing the social isolation of the elderly. Addressing social isolation was usually a by-product of the individuals delivering the service taking the time to connect with the client beyond their remit. In the more remote communities that have seen significant population decline, we were told of many instances of multiple single-elderly-person households, separated by empty properties. While some were still mobile and driving, others were less fortunate. And even for those that were still able to drive and get about, there was said to not really be anywhere to meet up. In the summer everywhere was too busy, and in the winter everything was closed.

> 'And some of them [older people living alone] do feel frightened if they're on their own at night and you're locking the door and there's a keypad on the outside, you put the key in and then they're there all night until you come back the next day. Yes. A lot of them do feel very isolated … the old people.' (Individual)

> 'And sometimes [as a home carer] you're the only person they see … you might be the only person they're seeing all day.' (Individual)

If older people are to be enabled to stay in their own homes through the provision of home care, their needs go beyond just daily care: they include having a warm and comfortable home, not being socially isolated and being able to manage their financial affairs and budget. Having a warm and comfortable home in old age can be a challenge in Harris' typically old croft houses. Finally, it is worth remembering that inadequate funding for social care in the community has consequences not only for the individual but also for NHS costs and waiting lists:

> 'they've got the biggest discharge, one of the biggest delay discharges out of hospital and that's because there's no home care, they can't get them out of the hospitals … And it's horrible for the people as well. Imagine being at the end of your life and you're trapped in hospital wasting your life in hospital which I know one of my husband's cousins is going through it and another person as well.' (Gatekeeper)

It was felt by some respondents that there was an impact on mental health of those that were isolated, regardless of their age. Part of this was the psychological effect of looking out on to a dark village in the winter, the lack of lights on in the houses a very visible reminder of what and who has been lost from the community: "so, there were times where, see there's 50 odd houses [in this township], 12 with lights on this winter" (Individual).

Isolation was one aspect of the 'hidden' life of Harris, and poverty was another, just as in Perthshire. There was general agreement that there were pockets of poverty in the community, but it was very hard to see because it wasn't concentrated in the same way that it was in other parts of the islands and on the mainland. The physical beauty of Harris, and the very evident wealth of some residents, were also said to help mask the more deprived circumstances of others. People were said to be very good at hiding their own poverty, due to the dual characteristics of a fear of stigma and a sense of pride.

'I think there's a perception on the island by some that there's not really poverty here, or that doesn't really exist on the island … Harris isn't just some affluent place where lots of rich people live – it's a normal place, but I think it does give you that impression. And they think there's not really anything … people seem to think there's no poverty here, or little, or they don't picture it in the same way as they would in an urban area and I think that's probably most true of Harris – people wouldn't expect people to be applying for a crisis grant living in Harris, but there's the same issues that are all over the island, getting childcare for your kids so you can go out and be active and work. Transport, access to amenities, etc, infrastructure, quite a lot of Harris is just a single-track road. Those issues are there but they're maybe more hidden than most people would be aware of.' (Gatekeeper)

A lot of community consultation has taken place in Harris to think about, and plan strategically for, future development. Central to this is the Outer Hebrides CPP, but also at a more local level the Harris Forum and HDL, coalescing around a pan-Harris Plan. As a result, many respondents had given the future of the island considerable thought and they had clear ideas about what they wanted to see. It was felt that there needed to be more recognition of the effect of sparsity on an understanding of the problems on the island: there is an 'averaging out' effect that hides the extremes of wealth and poverty, and therefore masks the small number of isolated but very extreme situations of some people in Harris. This was felt to influence decisions about policy and investment in many ways, from public bodies to third-sector funders.

'You have to be ticking so many different boxes – to get your funding … It's become almost impossible for a small community here to get funding because they don't fit in, you know, that criteria and there might be a criteria like how many single-parent families have you got? You know in South Harris you might find there are only two or three – and so for somebody … appraising the forms will say, oh they don't have many … It's difficult to fill in these forms … it's less than two thousand people in Harris now.' (Individual) 'the Scottish index, deprivation stats, even internally are only so useful and how many families do you actually have and what's the particular issues and how can you get easy markers that you can have a short questionnaire or just a couple of indicators that you go: "Right, okay, you're someone we have to go and talk to" ... So targeting is so, so difficult to get that right. The problem you have is your division between the haves and have nots is quite wide, so your average is not going to be an indicator, sometimes it's not a good indicator.' (Gatekeeper)

Conclusion

Harris has undergone dramatic change in the last half century, from decline to a fragile boom. Its remote location and small population make it vulnerable to the impacts of minor changes. Changes at a national and global level have resulted in changes at the community level, eroding the sense of community spirit, but the attachment to place is still very strong. Consequently, there is a strong sense of identity and a pull which draws people back to the island, and which has helped the development of the 'Harris brand'. Within this mix, individuals find their way, using a range of support mechanisms in daily life and also in times of trouble. These include informal support from family, neighbours and friends, as well as voluntary organisations (such as CABs, TIG, foodbanks and charities), community trusts, GPs and local government.

Unlike in many areas, financial vulnerability was likely to be the result of poor-quality employment rather than unemployment. And personal coping strategies were likely to be called on before resorting to state benefits, because of a culture of coping coupled with a sense of pride and feelings of stigma attached to more official forms of support. But for those that did reach out to the benefits system, the same issues of navigation of the system and coping with income volatility as noted in Perthshire were common. A specific island issue of restricted access to WCAs, and the impact of this on income, was highlighted.

Social isolation, as a result of, among other things, limited public transport, an ageing population and population decline, was increasingly a concern, particularly for the elderly, although young people were also in danger of being socially isolated. Social activities centred around community events and organisations, having the double benefit of providing opportunities for social contact as well as providing services. The key issue which impacted on all aspects of life in Harris, however, was access to affordable, suitable housing.

The North Tyne valley, Northumberland: a remote area of England

Like Harris, the Northumberland study is in a remote rural area, this time on the mainland of Britain in the north-east of England. Just south of the Scottish border, almost equidistant from the west and east coasts, the valley of the North Tyne river rises in the Cheviot Hills above Kielder Water, a reservoir surrounded by Kielder Forest. The largest villages are downstream, including the main settlement of Bellingham. The North Tyne valley study area includes the four civil parishes of Bellingham, Kielder, Falstone and Tarset and Greystead, covering around 530 km² (see Figure 5.1). The south-east part of the study area is within the boundaries of Northumberland National Park, which was designated in 1956. The National Park is the most northerly and most remote from large urban areas, least visited and least populated of the ten National Parks in England. The travel time by car to the city of Newcastle-upon-Tyne, which has a population of around 800,000, is 80 minutes from Kielder, or around 50 minutes from Bellingham.

There is a wealth of history in the area, with many scheduled monuments, listed buildings and archaeological sites. In the Middle Ages, this was one of the most dangerous parts of England and inhabitants had to live in a state of constant alert. Scotland and England were frequently at war and 'Border reivers' (raiders) along the Anglo-Scottish border robbed the entire county without regard to the nationality of their victims. Today, the beautiful valley is much more peaceful and home to just over 2,000 people – one of the lowest population densities in England. Most residents live in the Bellingham parish (1,325), followed by Tarset and Greystead (262), Falstone (245) and Kielder (187) (Office for National Statistics, 2018).

Although these four parishes are situated near one another, they are quite different. Tarset and Greystead parish and Kielder parish are two of the most sparsely populated parishes in the UK (1.5 people per km² in Tarset and Greystead), with ageing demographic profiles and a high proportion of older residents. The population has declined in recent decades in Kielder (-9 per cent between 2002 and 2017), a village developed by the Forestry Commission in the 1950s to accommodate the large number of workers

Figure 5.1: Map showing the North Tyne valley study area

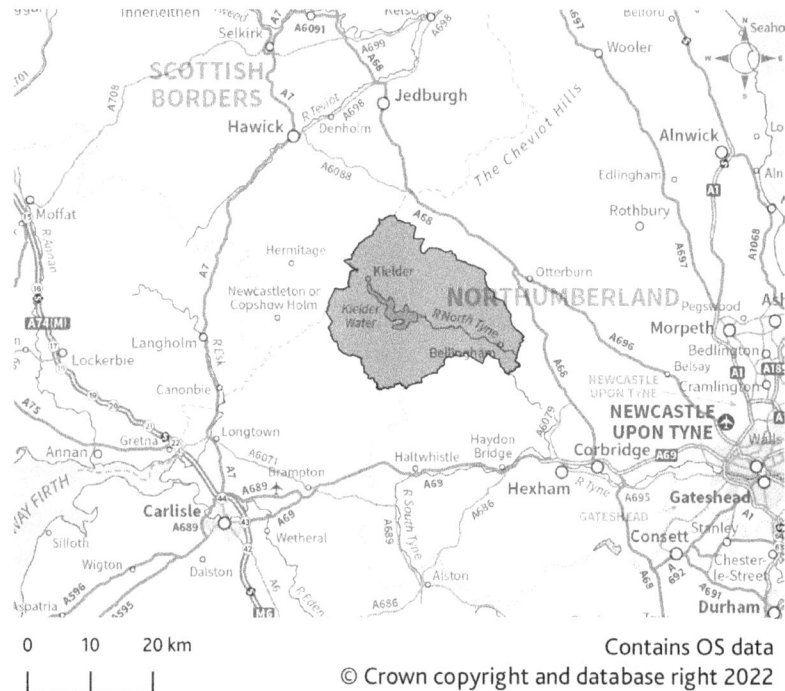

0   10   20 km

Contains OS data
© Crown copyright and database right 2022

expected to work in Kielder Forest. One interviewee described how "the working community has disappeared almost completely, and it has become almost totally holiday folk" (Gatekeeper) now that employment in forestry has decreased in the village. Bellingham, a popular stopping point for those visiting the National Park and Kielder Water and Forest Park, as well as those using the Pennine Way and national cycle routes, has seen a population increase (+4 per cent) over the same period, despite losing some local services in recent years. Tarset was described as having an extremely active community with "more IT professionals than farmers, and probably more artists than farmers" (Gatekeeper), with a growth in the number of "creatives" and those involved in craft activities in particular "taking off in the area" (Gatekeeper).

Key industries include hill farming and forestry, yet much less so than in the past. Although many more people would have had a connection to agriculture in previous decades, some young farmers remain, and farming succession was described as "looking quite good here" (Gatekeeper) and that "in general, around here, … certainly, the children are taking them over" (Gatekeeper). In all places, tourism and associated activities have become central to the rural economy in the last decade, particularly because of the

National Park and Kielder Water and Forest Park receiving England's first International Dark Sky Park designation in 2013. Despite this recent upturn in tourism, which has partly counteracted the decrease in agricultural and forestry jobs, one of our interviewees still described the area as having a continuing "history of loss" (Gatekeeper), which contrasts with the recent revival in Harris.

In the late 1950s the closure of the Border Counties Railway led to the loss of the train stations in Kielder, Falstone, Tarset, Bellingham and Redesmouth. Homes were also lost in Kielder, to enable the construction of the Kielder Water reservoir in the late 1970s and early 1980s. Foot and mouth disease had a significant impact on the area in 2001, including the closure of Bellingham's agricultural mart. Most recently, the COVID-19 pandemic has presented new challenges for the area, which are discussed in Chapter 6.

Like in Perthshire, the population is economically and socially diverse. Financial hardship is quite widespread in the area, with interviewees telling similar stories to those in Perthshire and Harris about their experiences with the welfare system. Like in Perthshire, there is also a disparity in wealth, with agencies not always having a clear picture about the existence and needs of rural residents experiencing poverty. Narratives of place and belonging are prominent, which allows analysis of how place modifies and intensifies the experiences and effects of financial hardship on age, gender, social relations and wellbeing in a rural setting.

Typical North Tyne valley landscape with small villages and isolated farms

Income from employment

Northumberland falls into the most deprived 20 per cent of local authorities in England for the number of people who are employment deprived and/or income deprived (Ministry of Housing, Communities & Local Government, 2019). The county is also in the top ten local authorities in England in the most deprived 10 per cent between the indicators of deprivation measured in 2015 and 2019 (Northumberland County Council, 2019), and in the top 12 UK local authorities that have seen a substantial increase in child poverty between 2014 and 2020 (Hirsch and Stone, 2021).[1] In 2018–19, Gross Value Added (GVA) per head in Northumberland was £15,564, compared with £20,554 across the north-east and £29,356 for the average across England (Northumberland County Council, 2020). This gap is much larger than for the Scottish case studies, when compared with the average GVA per head in Scotland. Total median full-time weekly earnings in Northumberland equated to £496 in 2019, compared with the English average of £591, although many rural residents do not work full-time so will not be captured in these figures. In the same year, the percentage of Northumbrian households classed as workless was 19.4 per cent, compared with an average across Great Britain of 14.3 per cent (Northumberland County Council, 2020).

The employment rate across Northumberland has been in steady decline for both men and women since mid-2012, following a particularly high period of employment (of over 80 per cent for men and 70 per cent for women) (Northumberland County Council, 2020). Recently, the employment rate for men has increased slightly (to approximately 75 per cent), while the women's employment rate has fallen to 68.3 per cent (Northumberland County Council, 2020). This is thought to be linked to the increase in the state pension age for women, which has affected the employment rate for older women.

In the North Tyne valley there has been a shift from the dominance of land-based jobs to tourism as the main source of employment – a feature common in many rural areas. Working from home has become more common, supported by the recent improvements in broadband connectivity, and there is a high dependency on seasonal work. The study area is within the most deprived 20 per cent of England in relation to 'Barriers to Housing and Services' and 'Living Environment Deprivation' (Ministry of Housing, Communities & Local Government, 2019). Approximately 7 per cent of the population in the valley is estimated to be experiencing deprivation relating to low income and employment. Financial vulnerability and financial hardship were perceived by several of our interviewees as quite widespread in the area, although less so in the more affluent Tarset and Greystead parish: "a significant proportion of the population would be struggling in all sorts of practical ways. It presents itself in varied ways" (Gatekeeper). Poverty was thought to be masked to some extent in the deprivation indices and other local authority data because much of Northumberland is prosperous. As one interviewee explained:

> 'we do have that disparity in the wealth – we've got a lot of people who are very wealthy and then a lot of people who have very little, and I think there is concern. I don't think we've got a clear picture on that.' (Gatekeeper)

We learned that financial hardship presents itself in different ways in the study area and that there are many people who do not have full-time, permanent employment. This makes unpredictable incomes a common feature of rural working life. In response, there is a tendency for people to have several jobs. For people in their 20s and 30s, for example, many of those who have not moved away to go to university or college or left because they cannot afford to rent or buy property in the area have multiple jobs in order to survive. It is common to combine bar work with mobile hairdressing, for example, or care work with DIY 'odd jobs'. For those in this age group with children, parents often work at different times of the day/evening to accommodate childcare needs (childcare options for younger children are very limited and often require access to a car). As one of our interviewees explained:

> 'The only way that a lot of the young families will manage to keep going is for one of the parents to be working during school hours … and then collecting the children from school and the other parent certainly working in the evenings. Lots and lots of them don't have permanent full-time employment although they are working full-time, and sometimes more than full-time and indeed not having a very full family life because there's always a subsection of the family that is getting together.' (Gatekeeper)

Many older people need to work into their 70s due to the increase in the state pension age, and lack of an occupational pension. This has also had an impact on the childcare options available to younger people because "a lot of parents in the past have used their parents as their main childcare provider" (Gatekeeper). Greater flexibility in working patterns was seen as a possible solution to this by one business owner we spoke to. They run a successful company which employs several local people who travel from across the study area (and further afield). Flexible working hours are an option for the staff so that they can accommodate childcare and other needs. A SureStart Children's Centre operates in Bellingham, offering a weekly selection of play sessions and other activities for parents and carers to attend with their children, rather than full-day childcare. However, one interviewee explained that there had been more comprehensive childcare options in the past:

'[Y]ears back there were family hubs around Northumberland and workers from different disciplines came together and decided which mix of them would be best to support a family. You've got the Early Help Hubs and Early Help Networks that are focused on under-4s, SureStart and children's centres, all of that, they're really struggling to provide – there's nothing like there would have been five or ten years ago.' (Gatekeeper)

We heard that "we get more people on zero-hours contracts … in the rural areas" (Gatekeeper) and that losing one job (even if you have more than one) can have disproportionate impacts because "it's actually [the] household that can lose out big time" (Individual) when the margins between income and outgoings are quite tight. Lots of people are paid less than minimum wage, particularly women who "are attempting to have a business of some sort or have been forced into self-employment because they're hairdressers or cleaners" (Gatekeeper). In these scenarios people may resort to casual work such as tree planting, stone walling or other odd jobs when needed:

A SureStart centre, attached to the school in Bellingham, provides vital children's services

'quite often you find jobs being done in the next village up or down where a person has just needed help and that's the unemployed or out of season worker.' (Individual)

While reluctant to note the role of the 'black economy' (cash in hand work), one interviewee pointed out that "this props everything up and avoids crisis" even though it is "daunting for people to admit they do this" (Gatekeeper).

'I suppose it's just the odd-job people that are probably going to be struggling a little bit. And a lot of that can boil down, in my mind … the farmers I work for are great guys, and a lot of them are very, very good friends of mine, but they can be very tight. You know, they will pay £40 an hour for a plumber, but they will only pay somebody working with the sheep £8 or £10 an hour or something. … I mean, they may be tight with money, but their hearts aren't. Their hearts are good. But then, I mean, it depends on the individual, what you are going to charge, and there's less and less people getting about. … But I would still say that they probably deserve more, because you don't get holiday pay, you've got your tax to pay and everything, and travelling.' (Individual)

Although there are still many young farmers in the area because succession of tenanted farms is strong, there are "lots of farmers very worried about future support [from agricultural subsidies]" (Individual) and many challenges experienced by 'loose farmers' – agricultural labourers who are neither tenants nor formally employed. They are paid for their work daily and can struggle to support themselves when work is either not available or they are not paid at the end of the working day. One interviewee recounted a story:

'I'll give one example that I did know about and that was somebody who was pretty desperate because it was Monday and he had finished all his groceries over the weekend, his week's supply. He'd done a full day's work as an agricultural labourer and he hadn't been paid at the end of Monday, so he actually had nothing to eat for Monday's supper.' (Gatekeeper)

Self-employment in the agricultural sector also presents challenges:

'There were times, in the early days, when I first went self-employed … and I went self-employed through redundancy. I was working … on a farm in the North Tyne. And there were no other full-time … jobs to be had, so I went self-employed. And in the early days, I received some sort of grant that somebody had put me forward for with the government, and you got about £35 or £40 a week. But I'm not ashamed to admit that I lived on bread and beans and things like that, you know, just to get through, sort of thing.' (Individual)

The requirements of casual work can also be unrealistic for those with other responsibilities:

'Quite a lot of casual or low-paid jobs are very demanding, they're either zero-hours or they're very demanding in terms of reliability. [For example], there's a sign that goes up in the garage … saying: "We're looking for staff, you have to be available 24/7". You can't work 24/7, you're not going to get all that money, but you have to be available, even more available than you do at the Job Centre, so that's awful.' (Gatekeeper)

The public sector employs a lot of women across Northumberland although:

'80 per cent of the cuts in the public sector centre [in recent years] were on women's jobs. That was a main career route for women in Northumberland, the public sector, in all its forms and they've borne a huge brunt of the cuts there.' (Gatekeeper)

There are also disparities in income between men and women, with one interviewee describing the "massive glass ceiling for women's enterprise" across rural Northumberland (Gatekeeper). Those businesses with a turnover of more than £20,000 per annum tend to be owned by men, with women's enterprise in general sitting below that threshold. However, "rural women [are] doing well in micro-enterprise" (Gatekeeper), despite the observation that many people in rural areas, particularly women, can feel very daunted by the prospect of self-employment because they misunderstand what is involved:

'There's almost no specialist support around that, we've tried to fill that gap, we run a course for becoming self-employed to try and take them through the implications of it a bit more. It's a big deal if you're coming off benefits to go on the enterprise allowance and it only lasts for six months and it's no more money and the idea of it is that if you start to trade in that time then you can keep any profits until the end of the six months, by which time you need to be building up and making a profit.' (Gatekeeper)

This interviewee also noted the need to improve rural women's understandings of 'small is good', which is challenging when a lot of the available start-up and other funding is targeted at business growth:

'You can contribute just as much to the economy by staying small if you want to but you can't access most of the grants or loans if you're not prepared to say you're going to double in size in the next year. There's huge issues around that. … A lot of people don't want to have a workforce, they don't want to become managers of what they do, they're creative people so they want to do it themselves. So that's another issue.' (Gatekeeper)

People who have multiple jobs were generally thought not to have any further or higher education qualifications. This presents employability and skills issues that limit the potential for people to enter higher-paid jobs. Although we were told that there are some new businesses in Bellingham which require employees, and that there is growing potential that self-employed workers in the area may seek to take on new staff, most of the people in financial hardship were thought not to have the technical skills that would be required in these roles. An attempt to develop such skills through an apprenticeship scheme was frustrated by the requirement for apprentices in Bellingham to find their way to the Team Valley (south of Newcastle) for their one day a week of further education (Gatekeeper).

Digital skills were also regarded as important when considering how to support people to apply for higher-paid roles although literacy issues may also present challenges:

'Most of the people that we are worried about with low income in the rural area do not have [a] good enough educational background to go into technical employment. … You've got an awful lot of obstacles to start to get these people into some sort of … training preparing for any sort of work.' (Gatekeeper)

It can be the case that some individuals are not able to work at a level suited to their capacity/ability, particularly women who have caring responsibilities. There are also challenges associated with sourcing suitable care for those who need it (for example, we heard an account of a vulnerable teenager who was not able to get the care they needed due to unavailability of suitable care support in the rural area). As one interviewee explained:

'We've found a lot of women are pretty well educated in terms of their schooling a long time ago, but they haven't been in work. … There's a lot of under-capacity of really talented women because they're taking on such complex caring roles.' (Gatekeeper)

Lower-end jobs have either been mechanised or the distance from potential large employers makes travelling to work unaffordable. There are haulage/

buildings contractors about ten miles from Bellingham but no public transport routes to reach them, and the use of contractors from outside the area is common:

> 'You've got people repairing roads, laying tarmac [and] putting in service ducts for broadband. But many of those things are done by contractors who come in from elsewhere. So ... they're using trusted workers that they're bringing from other parts of the country to do a job here and then they're going. So, there's no long-term steady methods of working that's achievable for these people.' (Gatekeeper)

Many of the area's population are beyond commuting distance to the nearest larger town of Hexham (40 miles from Kielder, 25 miles from Falstone), with "distance to work" (Gatekeeper) amplifying these employability challenges. Petrol and diesel are on average 30 pence more expensive per litre in Bellingham than in Hexham, although the unmanned petrol station in Kielder is more competitive. Travel costs contribute to in-work poverty, with the distance/cost to work being one of the main reasons that people seek loans from the local community bank.

Support from the state

Across Northumberland, 3.3 per cent of residents aged 16 to 64 claimed out of work benefits in 2019. The claimant rate in the North Tyne valley is estimated to be between 0.5 per cent and 2 per cent (Northumberland County Council, 2020). We heard about similar issues in Northumberland to those we learned about in our other case study areas. For example, the complexity of the system, Universal Credit (UC) payment delays and challenges associated with the Job Centre, Citizens Advice Bureau (CAB) and medical assessment centres being located far from home. Our interviewees described how being a rural claimant compounds all the issues with the system. For example, claimants must travel to Gosforth in Newcastle for medical assessments (53 miles from Kielder and 34 from Bellingham), with no suitable public transport available to make this trip from north of Bellingham. This can make it unappealing for people to claim the support they are entitled to.

In West Northumberland as a whole, it is now more common than in the past that people reach out for support for navigating the benefits system. Two of our interviewees highlighted that benefits advice is now "the biggest enquiry issue that we deal with" (Gatekeeper), rather than debt, which was the most common reason people would seek advice in the past. Of the reasons people approach the CAB in Hexham for benefits advice, Personal Independence Payment support is the most common, followed by UC initial claim support and Employment Support Allowance (ESA) support (we learned this from our interviewees and the Hexham CAB key statistics, 2019–20). These observations confirm that "it's definitely the sickness-related benefits that we're seeing more people with than anything else" (Gatekeeper). It is notable that 50 per cent of the Hexham CAB clients have a disability/long-term health condition and lots have literacy problems: "there is a very large percentage who can't read or write" (Gatekeeper). More generally, we heard how advice provided by the CAB and others in the past was on "how to maximise benefits", whereas now the focus is on "how to survive with very little" (Gatekeeper).

We learned again how the benefits system is "very complicated for the most vulnerable people" and that it can be "quite distressing" (Gatekeeper) for some people. Concerns were raised that the complexity of the system has led to "people … getting turned down … without actually looking into, properly, their health issues" (Gatekeeper). Noting that many claimants are also turned down at the appeals stage, it was concerning to hear that "[t]hey don't understand … why they're being turned down at all. It's fairly obvious that they're entitled to the benefit a lot of the time" (Gatekeeper). Appeals can also take time, "more than a year sometimes, from the submission of the appeal to actually being heard" (Gatekeeper) and there is an increased prevalence of UK Department of Work and Pensions (DWP) representatives being present at the hearings. Like in Perth and Harris, we also heard stories of the unkindness and emotional distress that individuals experienced during the claims process:

'But they made us feel horrible about trying to claim a benefit. I didn't want to, it's just I need to. I was ill at the time. I was really bad when I went in. And I said, "Look, I'm suffering from really bad headaches. I suffer from chronic migraines. There's the letter. There's the medication I'm on". They look at me like I'm just trying to put them up to answer the questions on whether I can go on to benefits. I had to go through a phone call to a different part of the country, because I was just in Hexham, to make the claim. And they wouldn't do it in Hexham. They wouldn't help us. So, I got on the phone, they dialled the number, and I had to tell them I had been in, and then they got processed, I think she said, it took about 14 days but everything was paid backdated. But it was just so, they made us feel so small and there was no help at all.' (Individual)

Concerns were raised about the switch to the new UC system, particularly for those who are long-term claimants, don't have good IT skills and/or don't have easy access to a device to log in and update the online journal. As one interviewee noted:

'you're talking about people who have been on ESA and PIP [Personal Independence Payment] for a lot of years. Transferring them over on to Universal Credit is going to be catastrophic for some people. It's going to be a very different way for them to qualify for their money. … Obviously, there's the online way that you have to claim it which, for a lot of people who haven't been in the workplace for a long time, they don't have the IT skills, they don't have the IT equipment. Then, of course, in the rural areas, you've got the added issue of broadband, access to get to the libraries where there may or may not be computers and people to help.' (Gatekeeper)

Although there is a public library in Bellingham, which is open on two days each week, we were told that there is unlikely to be anyone at the library who can support claimants to complete the online journal. This type of assistance is available in some of the larger towns in the county (for example, Ashington and Blyth). However, we were also told that, when people in the study area need to update their online journal, they are generally quite able to access the nearest hub (perhaps by phone) or get online to complete an entry.

Despite the concerns raised about digitalisation, these were less significant in Northumberland than in the Scottish case studies, perhaps reflecting the more reliable broadband infrastructure in the study area (93.6 per cent of properties in Northumberland have access to superfast broadband.[2] However, concerns were raised in one focus group that the benefits of improved digital infrastructure are not reaching all who need it.

'Whilst the county have done a wonderful job of getting superfast broadband in place, actually poor people don't have it, so it is massively a problem. They don't have it on their phones, they don't have data. They don't have the signal strength on their phones to be able to access things remotely, through their smart phones, and they don't have a paid-for provider at home because they can't afford it.' (Focus group)

Rural people were described as having adapted well to managing their UC accounts via the phone or at the Job Centre if they are unable to get online, with the concerns about digital access not having played out as badly as expected.

'certainly, digital access is still an issue, but I still think there's a lot of working-age people with enough technology, mobile phone technology, to manage Universal Credit accounts themselves. There's still big gaps in terms of access to digital services but it hasn't been as cataclysmic, in terms of people being disadvantaged, in the sense of being sanctioned because they haven't been able to get to a digital hub or to access their claims online, in time to action all the work instructions. That might go on but that's not necessarily what we've seen in terms of the cases that we're dealing with.' (Gatekeeper)

It was, however, noted that digital exclusion may present a hurdle for older clients, which is important because the majority of CAB clients are currently 50 to 69 years old and there has recently been a rising prevalence of requests for support from people in this and older age groups. Broadband connection speeds could also still be improved in some places and mobile phone signal is very poor and, in several places, non-existent, which is felt to add to social isolation.

We heard again how the volatile nature of rural incomes means that claimants cannot plan/budget easily as they do not know how much income support they will receive each month if their monthly income changes or fluctuates.

> 'A lot of that is because Universal Credit is designed based on a system of monthly salaries and people who are working full-time. The reason why I mention is because the employment demographic that you've detailed at the lower end of this, Universal Credit is a nightmare. It is apportioned over 12 months when, obviously if you're paid weekly, or week to week, it doesn't work. They end up with two payments in the same month which has a massive knock-on effect and they know about it and they haven't done anything about it yet, or are likely to. It's those people at that end who are really struggling with that.' (Focus group)

As one interviewee noted, those who have several jobs have an "incredibly flawed relationship with the benefits system because tracking the income from the jobs is very complex" (Gatekeeper) and over/underpayments and the need for advance payments create additional issues:

> 'obviously if people apply for an advance on their Universal Credit, they do have to pay that back and then there are some people actually who I've spoken to have had an advance and because of the fact that that's been deducted from their payments, it's kind of leaving them really quite short in terms of being able to manage their kind of monthly budget.' (Gatekeeper)

This is compounded by the fact that the day of the month on which an individual receives their wages can also change, in turn affecting how much support can be claimed within a given period and/or lead to a delayed payment which can leave people "quite desperate" if they are not receiving payments for "five weeks or so" (Gatekeeper):

'Depending on what they earn, it's very difficult for them to plan and budget. … [P]eople are finding it hard to change the date of the payment for Universal Credit … which would be the sensible thing to do but then, the system doesn't seem able to do that. … But then either the employer can't change the time that they pay them, and the Universal Credit can't change the date of that payment, so they're stuck.' (Gatekeeper)

A common issue in the rural employment scenarios outlined earlier in this chapter is that people can quite suddenly find themselves out of employment with no formal letter/exit process from their employer. This can also lead to payment delays and challenges associated with registering income in the system. Payment delays can lead people to "approach loan sharks because that's the only way that they can see getting through this period" (Gatekeeper).

Payment delays can also have greater negative impact in the autumn/winter when people's energy bills are higher, "particularly from about October last year when the temperature started to drop then we saw a huge influx of referrals for people needing emergency fuel top-ups because … they kind of applied for Universal Credit and were struggling to budget in the meantime while they wait for their first payment to come through" (Gatekeeper).

In contrast to the Scottish case studies, "there isn't the option [with UC] to pay the Social Housing Landlord direct by request … only if there's a need for a managed payment, or an alternative payment arrangement, because of rent arrears or vulnerability and so on" (Gatekeeper). This has led to rent arrears and presents housing associations and other housing providers with the challenge of managing those arrears.

Overall, one interviewee thought that "there are not many people within this area that are yet to rely on a state pension", although he thought this may change when the "generation who aren't doing forestry" are no longer here (Gatekeeper). There may be stigma attached to claiming benefit entitlements, particularly the attendance allowance (described as "the most under-claimed benefit" [Gatekeeper]) and carer's allowance, although there was no official data available for the study area on whether people are under-claiming elements. One interviewee suggested that older people may lack awareness of entitlement to Pension Credit. These issues may be linked to the "private" or "proud" nature of people in the study area and the common situation where "everybody thinks that somebody else is worse off than them" (Focus group).

'We had a lady in the other day who was entitled to carer's allowance because her husband was ill. She was arguing that she wouldn't claim carer's allowance because she was his wife and she cared for him anyway.

We were trying to explain but she was adamant she didn't want to claim it.' (Gatekeeper)

The language used to engage people who may need support was also mentioned in some of the interviews. Particularly as a response to the "proud" attitudes of West Northumbrians, how help/support is framed was seen as important for encouraging uptake of advice and support.

'it's the wording. It's about how it's sold. There is this stigma which I think is fuelled by the media about claiming benefits, whereas in fact, it's people's rights. It's just there for people to claim, it's there to help them manage their lives so their finance is better. So, I think there's probably something about the language that's used, particularly with the older generation as well. I think if it can be tied up with their, we can then say, "Come along to a tea and coffee meeting, and oh, by the way, do you know about attendance allowance?" I think that's the way to really talk to people about it.' (Gatekeeper)

Despite these challenges, another interviewee noted the effectiveness of the carer's allowance in a rural setting in the following example:

'Mother and father were in the village, son and daughter were there. They were giving up their work to then look after mother and father by getting carer's allowance and then doing seasonal work that would fit the 16 hours available. So, they focused on the family. Quite often, mother and father, if they were pensionable, would be sharing the family money together. Does that make sense? … The system actually worked for what it was meant to do which was really good.' (Individual)

When discussing how to raise awareness of these types of state support, it became clear that how far agencies can 'reach' into rural areas is key.

'We've got a high level of people who are paid below minimum wage, we've got people who we know are entitled to benefits, who don't access benefits, who don't access health services and we rely, a lot of the time, putting information in health places. So, that reach just isn't there [for the rural areas].' (Gatekeeper)

One interviewee also raised the concern that:

'it is still those people who are the furthest removed, the people who are the hardest to reach or who are most isolated are the most vulnerable.

They're the ones who are continuing to get left further and further behind as everything else progresses.' (Gatekeeper)

'Communication is a massive thing because you've got mobile phones that are working. You've got talking to farming communities when they're out of the house the majority of the day. So, you can't catch them on the landline. You know as well as I do, just that whole communication and actually that's a massive priority. How do we let them know that we're here? How do we let them know what's available to them?' (Focus group)

One potential solution for agencies lies with the local care coordinators employed through the Northumberland Health Trust. For example, some organisations have "put leaflets within the care coordinators … all we can do is signpost people and the word 'signpost' is more of an acceptable word to use than refer people" (Gatekeeper). A local housing association also offers a unique benefits advice service to its tenants and the number of people being referred to the service "increases all the time" (Gatekeeper).

There appears to be renewed effort from Northumberland County Council to improve the organisation's reach into rural Northumberland, recognising that rural residents can "quite often … become invisible and hidden", particularly as Northumberland is quite a "wealthy, well-performing county generally" (Gatekeeper). The CAB has also been working hard, despite funding constraints, to extend their reach into rural Northumberland, but this is not straightforward and providing a face-to-face service presents a real challenge.

'because we are a large, normal Citizens Advice, we have to deliver in a very different way to city centre offices. … We haven't been allocated a resource which takes account of the rural nature. It's just been allocated on the number of people anticipated that will be claiming Universal Credit. So that makes it really difficult for us to provide a face-to-face service across the whole of the county on the limited resources that we have. So yes, we will help people up until their first correct payment has been placed. Anything that's beyond that, they've had to go into our general service which again then comes down to resources and we're going to be able to help people on the limited number of people and volunteers that we have ready to help people.' (Gatekeeper)

The CAB receives approximately one third of its funding from the local authority, one third from the Money Advice/Debt Advice Service and the remainder from different project funding and donations from town councils, for example. Although funding from the local authority has not decreased in recent years, which is not the case for all CABs, funding constraints limit how much they can achieve, particularly in the current scenario of rapidly rising demand for their services. A couple of years ago, the decision was made to reduce their face-to-face services and put more resource into the CAB telephone service. Despite the availability of an enhanced phone service meaning that clients based in remote parts of the county can access advice without needing to travel, concerns were raised about vulnerable people who are no longer able to access face-to-face support.

> 'it does mean that those who can access a telephone and use the telephone, if they live in the middle of nowhere, then they have more chance to get in touch. There's more likelihood to be able to help them than rather than get into the Hexham office, for example. But it does mean that those really, really vulnerable people who need that face-to-face support, you know, that support just isn't there as much as it used to be. [The] Hexham office is now only open one day a week.' (Gatekeeper)

With no designated outreach visits/drop-in services in villages in the study area, the CAB tries to link up with other organisations working face to face with people in rural areas. By doing so, they try to ensure that other organisations are aware of their services so that organisations can refer those who would benefit from their advice. CAB drop-in sessions remain in place in Hexham on Fridays.

> '[We are] trying to link up with those organisations and pass on the message that if you can help to identify some of these issues when you're going into somebody's house and you kind of think, well, this might be a warning sign. It's just … simple awareness but it could mean that people could ask another few questions while they're there and maybe help to link that person into support to be able to kind of assist with maybe fuel poverty, that sort of thing. [We've done] sessions with housing associations and with [the] voluntary sector … community groups … and that sort of thing.' (Gatekeeper)

Despite these challenges, the CAB has also been running an energy project to tackle fuel poverty and digital exclusion issues. Fuel poverty is a significant issue in rural Northumberland: advisers working on the project carry out a household assessment to work out consumption, property type, energy efficiency and so on. They then "look at what the costs are, the payments, any vulnerabilities and things like that. And we combine all of the information that we gather at the start to offer a support package to people" (Gatekeeper).

"Because people who can't switch because they don't have access to the internet, then the fuel workers can actually help them switch and go to the dual system, so I think that's helpful for them" (Gatekeeper).

Reaching those most in need of support was also a prominent concern of those working on the energy project:

'what we tend to find is that the people who may be most at risk in terms of fuel poverty in the county tend to be people that are hardest to reach.' (Gatekeeper)

Language is also important in this work when advisers are discussing energy saving measures with people:

'we don't mention the word fuel poverty, because people don't want to admit that they're in fuel poverty, so we do look at other ways of encouraging people to think about it. So … putting a more psychological slant on it, about saving the planet rather than about saving the actual individual the money.' (Gatekeeper)

One of our interviewees also suggested that while residents might be seen nationally as experiencing fuel poverty, local people may not see themselves as poor. This may also contribute to some people not claiming the benefits they are entitled to.

Support from voluntary and community organisations

It was striking how many third-sector organisations are offering support and advice in the study area and across rural Northumberland, despite the difficulties associated with 'reach' by the state into rural areas outlined earlier. We heard about the proactive attitudes of people in the west of the county and how they have decided they have to "do it for themselves" if they want something to happen.

The former local authority (Tynedale District Council) supported the establishment and operation of several Community Development Trusts (CDTs) across Northumberland before its closure in 2009, including the Hexham Community Partnership, as well as CDTs in Bellingham and in Kielder. Today, the remaining Trusts are still seen as "trying to gap-fill to a lesser or greater extent" (Gatekeeper) between public and private sector provision.

Our interviewees described many examples of the positive and enterprising work of voluntary and community organisations in the study area. For example, the North Tyne and Redesdale Community Partnership (NTRCP), which is wholly run by volunteers, has done a lot of work to address social isolation (organising events, running trips and so on) and runs a youth group in Bellingham. The Trust also manages a business premises in Bellingham and is in the process of developing four new office spaces in the village. This is in recognition of the lack of start-up office space, which limits the potential to create new employment opportunities in the area. The initiative has proven popular:

> 'we certainly don't have any problem letting the office space that we already have. And we have a waiting listing for people that want office space.' (Gatekeeper)

Kielder Ltd, a social enterprise described by one of our interviewees as "the second-best development trust we have in the county" (Gatekeeper), runs an unmanned fuel station, campsite and local shop in Kielder village. Tarset was repeatedly described as having very strong community infrastructure and networks and a draft Neighbourhood Plan for 2015–35 was recently drawn up at the initiative of local people in the parish, although subsequently withdrawn.[3]

Community Action Northumberland (CAN) works across the county to sustain and support rural communities via a range of initiatives. Of particular note is their county-wide network of 26 'Warm Hubs', which are in local community places (usually village halls) where people can find a 'safe, warm and friendly environment in which to enjoy refreshments, social activity, information and advice and the company of other people'.[4] Visitors to the independent, volunteer-run hubs can access energy audits and advice about energy efficiency and other support, and the service reaches out particularly to older, isolated people and parents with young children. The nearest hubs to the study area are in Hexham and Haltwhistle, providing places

> 'to informally put messages out, help people change supplier and those sorts of things and then often on the back of that we'll have some of the health service people wanting to come and have a chat with people while they're there about different things. People are coming because they want to socialise and have a meal and have a good time while they're there – we informally do the other stuff and let our services come along now and again and talk to people, but it's very informal. It works really well.' (Gatekeeper)

The Warm Hubs project receives funding from an energy supplier and CAN helps the hubs to access additional funding, for example for upgrading village hall facilities and for training hub volunteers (for example, dementia training and benefits advice training). The volunteers were described as: "local feet on the ground to try and encourage people to join … word of mouth has always been, and remains, the most effective communication mechanism when you get to very rural areas" (Gatekeeper). CAN also runs an Employment Hub once a week in Hexham to support unemployed residents with their CV and job search, providing a 'next step' from the Job Centre.[5] We learned that, for the people using the employment hubs, common issues include isolation and mental health issues, with many people having multiple vulnerabilities.

'The client group we get, some of them don't need much help, but an awful lot of them need all sorts of help. Yes, they're out of work, they might have been out of work for some time, they might have been caring for somebody, they might have had some life incident or whatever, they're very isolated, they're very lacking in confidence, they've got mental health issues and it's this whole gamut of things they're suffering from that we try and help them through with the employment bit as the focus. [We also offer] slow cooker courses, all sorts of stuff, wellbeing services. There's lots of things we do beyond the employment help.' (Gatekeeper)

More broadly, CAN provides support for village halls across the county, as part of the Action with Communities in Rural England (ACRE) network of rural community councils, which "supports village halls across the country, supporting them with good governance, their improvements, refurbishments, new builds, all sorts of stuff, queries and they are places that stuff happens in rural communities, so they're really important to us" (Gatekeeper).

The West Northumberland foodbank is in Hexham. One interviewee explained that "we've definitely got people in Bellingham and the surrounding district using the foodbank and have had for quite a long time" (Gatekeeper). Most visitors to the foodbank need to access the service because of mental and/or physical health problems (over 40 per cent of users in 2017–18). Other reasons include debt/budgeting issues, benefit payment delays, homelessness and low pay. 94 per cent of visitors to the foodbank claim benefits of some kind, at least 14 per cent are employed and over 10 per cent attend due to hardship caused by delays in processing benefits claims (West Northumberland Food Bank, 2021). It is worth noting that users of this foodbank do not require formal referral to be able to use it and that the staff are "doing all sorts of stuff that go beyond food ... they're doing finances" (Gatekeeper). There are two drop-off points in the North Tyne study area for donations to the foodbank, but those in need of its help generally have to travel by bus to Hexham to collect provisions from the foodbank.

'They have a team of what they call support workers who do financial support and also, they have ... a no criteria approach. ... So, people come once a week and they accept that their income is not enough, so they just support them with a food parcel.' (Gatekeeper)

The foodbank also provides a place for people to talk, particularly when someone is "not really that open with people that I know" (Individual) and talking to strangers going through similar experiences provides a sense of relief.

'You just sit and you talk about your problems. It's kind of a, it's stress relief. It's kind of, it helps you for that day. It helps you get through it. I know it's an 18-mile or 36-miles round trip to Hexham and it was costing me £5.20 on the bus. [But] it's so helpful – they are really great.' (Individual)

We heard that the community-run jumble sales in Bellingham Town Hall are very important, both in providing social events and the opportunity to buy high-quality items at a low price. One interviewee described these sales as one of the available "strategies for surviving", albeit an informal one:

'Jumble sales are very popular. ... People go to a jumble sale or anything in aid of charity with a £10 note and when it's spent, it's spent and that's it. I think that's how quite a lot of people manage to buy things that they would like because they're all quite high-quality things, they're not rubbish. So, they are other people's unwanted things but they're actually things that somebody else might be glad to have. I think that probably is one of the reasons why it works so well and an ongoing custom in Bellingham is that that is how people are managing to buy things other than food, even big pieces of furniture sometimes.' (Gatekeeper)

Despite all of this work by voluntary and community organisations, several interviewees raised concerns that the reduction in core funding from government for voluntary and community organisations has led to aggressive competition more broadly between charities to access limited funds. These concerns are a reflection of 'roll-out' neoliberalisation described in Chapter 1. Another common impact has been that these organisations are preoccupied under austerity with ensuring their financial sustainability and, as a result, are often less able to focus on the real needs of the communities that they serve because "if you start focusing in on your own sustainability too much, you forget the reason why you're there in the first place" (Gatekeeper).

In some cases, increasing pressure on trustees and volunteers has also made it challenging to have enough people to sustain the work of the organisation – we heard that this is a common issue across development trusts in the county and can contribute to the lack of services in some places. This is the case for Kielder Ltd, which currently has only four trustees, the minimum to be quorate, which makes its operations vulnerable to closure.

The viability of setting up new businesses/services has presented a challenge for some voluntary and community organisations. For example, the NTRCP had the ambition to reinstate employability/skills support in Bellingham, to address some of the issues related to rural employment that were outlined earlier in this chapter. However, this has not yet been possible because although they "have some good core people that could be turned to for advice", they "couldn't staff it" (Gatekeeper).

CAN experienced similar challenges when considering establishing an Employment Hub in Bellingham, again related to the difficulties of 'reaching' into rural areas to offer support.

'the numbers were so small, like one or two people, we couldn't sustain it. … [People from] Bellingham are coming down to Hexham and we're helping with the travel costs of getting them there, so we've tried to do that. But we, ourselves, who are very rurally minded, we struggle to reach the really rural places and I'm always mindful of that in everything we do.' (Gatekeeper)

The cost of providing care in rural areas was also discussed as a barrier for the voluntary and community sector in this context, causing some local charities to "disappear" (Gatekeeper). This is often because

'delivering the same service to less people over a big area costs more per head, completely unrecognised in the government formula for our case and for local authorities – they know it's not recognised but they've not been able to do anything about it and it's just got worse and worse and worse.' (Gatekeeper)

Responding to these challenges and recognising the importance of offering joined-up advice and support, several of the interviewees noted how voluntary and community organisations are integrated in the way they offer advice and/or refer people to other services. These comments were made by both those people directly involved with voluntary and community organisations and those observing their work from afar.

'some people may qualify for foodbank vouchers, may qualify for funding that's available, may qualify for mental health services. There's a whole host of other services. It may not be financial help that they need. It may be additional things. … The more we are integrated with these other services we can help the end-to-end journey of those individuals who may be vulnerable in their rural areas.' (Gatekeeper)

The various voluntary and community organisation initiatives have also led to the development of an informal network of community volunteers who themselves bring another layer of connectedness to the support available to rural people experiencing financial hardship. One interviewee noted the potential for agencies and other organisations to link up with these individuals to build trust and help address the challenges of 'reaching' into rural areas.

'there are people within the smaller kind of villages and towns that, they're kind of volunteers in the community and they bring a lot of people together and sit on a lot of the local networks and things like that to make sure that the information gets back to people in their local community. And it's great, where we can find those people, tap into their sort of knowledge, because it's a trust thing as well. It's hard to kind of just go into any smaller communities and say, "Right, this is the information that I want to give you". It takes quite a lot of time and commitment to build up that trust with smaller communities especially, that might be quite closed off.' (Gatekeeper)

Being able to get support from trusted individuals within rural communities was also seen as very important. One voluntary and community sector representative described the "enormity" of the digital barriers associated with claiming state support and how individuals need help from trusted sources like her:

'the help to fill in forms, I mean, I have had people arrive on my doorstep over the years, knock on the door and say, "I just can't understand how to fill this in". That's alright though. I mean, that's the sort of thing I'm here for.' (Gatekeeper)

However, concerns were raised about ensuring that the quality of advice given locally meets the standards required by the CAB and others.

'So much of it comes down to the proper advice that someone gets when they're making the claim. Too many people think that they can help people to put their Universal Credit claim in but actually unless they have everything nailed down, they know exactly what those dates are, they have all their proof of eligibility and all of that sort of stuff ready to go, it's really counterproductive to do that. That's one of the things we try and get across to people. It's all done from a really well-meaning place but what tends to happen is because people understand the technology of applying for it, they just help people to do the actual application, rather than having that independent holistic advice as to whether it's the right thing to do for that person at that time. … It's increasingly frustrating for us that lots of people are trying to help, but not necessarily helping people in the end.' (Focus group)

Support from friends and family

We heard less about this type of support in the North Tyne valley than in our other case studies. However, community spirit appeared strong in parts of the study area. In Tarset and Greystead, people were described as "aware of each other and what they need" and "in and out of each other's houses" (Gatekeeper), routinely picking up shopping for other village residents.

> 'I feel more that, with the community that I'm most heavily involved in, people would notice if there was an issue. And actually, there are people within that community that would offer that help, if they thought it was necessary.' (Gatekeeper)

Bellingham was described as having "a great local tradition of accepting all of the community, accepting everybody for who they are" and that "people would notice if there was an issue" (Gatekeeper). One of our interviewees told us that people in Bellingham with mental and/or physical health issues are likely to live with supportive friends or family, rather than alone.

> 'We all know, by first name, those people who might be walking about the streets who if they need to be guided home or anything, we would know them by name and look after them as our neighbours. So, it seems to me maybe that people are able to just stay in their own community but, of course, they're not going to be very affluent and probably their families aren't. … I don't think the sort of people I'm implying would be living on their own. They would have family members who would be able to see about vital things that they might need extra help with.' (Gatekeeper)

In Kielder one interviewee described "a friendliness which is very pleasant, an openness in that some people will ask someone else for help [or] someone will know things because someone else has asked" (Individual). A specific example was the existence of trusted groups within the community, which made it easier for people to access informal help at challenging times.

'You can get things on account so to speak. We still get the coal man. He knows that you own your house – you're not going to do a midnight flit. He'll just come back to you and just pick up the money later on. … Garages, some of them are more accommodating than others, local garages. That helps oil the wheels of difficult times.' (Individual)

Within the farming community, although "you would never admit to anybody that you were hard up", people "would often … you know, if they killed a lamb or anything … they would give you two or three chops. And some folks would give you a load of logs or something like that. You never asked for it. You never let on that you required anything like that, but they would just do it" (Individual).

People in the west of Northumberland were repeatedly described as being quite private, with a "we're fine, thank you" attitude. While this demonstrates their general resilience and competence, we heard how this can also make disadvantage hidden. For those not "keyed into the local grapevine" (Gatekeeper), it can be hard to access local support networks. However, this privacy seemed different to the "stigma" associated with needing help that we encountered in the other study areas. Instead, it might be those people who are less well integrated into the communities in the study area who may struggle to ask for or access the help they need. Older people generally have an established social network, although leaving the study area is a challenge due to the lack of public transport. The area is a popular place for older couples to retire to after their working life and with a private pension. However, in some instances when one person passes away, the widow/widower could experience social isolation if they had not made the effort to integrate into the community when they first arrived.

'We tend to think about poverty among the young but it's the elderly who are suffering on the state pension. Those from high-pressure jobs retire here for peace and quiet with extra bedrooms for visitors – one partner dies then they are living in a big house and are forgotten about because they didn't make the effort to integrate when they moved in.' (Gatekeeper)

The privacy that rural living offers can act as a "pull factor" for some people, including families with children with disabilities who "choose to live very rurally because they have tough times with neighbours when they've lived in a street and people are looking at their child's meltdowns and the pressure of trying to keep your family neighbour-friendly becomes too much" (Gatekeeper).

More broadly in rural Northumberland, it may be that some people struggle to ask for help from their own family:

'I've spoken to a lot of people who've said that even members of their own family haven't owned up to the fact that they're struggling.' (Gatekeeper)

We heard a story about an individual who "didn't even realise that [his] own parents were struggling until [he] went one day and saw that they were kind of sitting with their coats and then I flicked the heating on as I walked in" (Gatekeeper). One individual explained the sense of pride in more detail:

'We're a very proud breed. We're not going to go and tell people we are strapped for cash or we've got problems. I don't know that I can. I do know of one guy who fell on bad times, and people were giving him furniture and such like, and making sure he got a hot dinner and things like that. We presumed that he would be strapped for cash. He didn't have any furniture in the digs he managed to get. And, you know, we looked out for them. But I don't recall that he ever said as much or asked for any help. But you just, sort of … you know, "Do you want to come for supper? I've got a spare bed that I was going to burn. Is it any use to you?" and that sort of thing. I mean, unless they come forward and specifically ask, then you've just got to try and read between the lines and see if you can help without … you know, and maybe ask them if they want to do a day's gardening for you because you haven't got the time or something like that.' (Individual)

One of our interviewees also suggested that young people in the area experience "poverty of contact", which leads to a more proud/self-reliant approach to life.

'Kids live in such outlying areas that they don't get to mix with their peers as often as kids in towns. You know, kids in towns are always saying there's nothing to do, but if you amplify that by being five miles from your nearest neighbour, then you have to be much more self-reliant. And I think that's what it does – it brings up kids that are much more self-resilient and reliant than you'd maybe find in urban areas.' (Gatekeeper)

Single mothers raising their children in rural areas often experience additional challenges if they do not have support from friends and family locally and wish to (re)enter the labour market. Inadequate housing, transport and communications infrastructure also compounds the isolation of rural women (Women's Resource Centre, nd). As one interviewee explained:

'if a single parent wants to work, say a woman wanted to try and develop a career, so she might have to travel to Newcastle to work and she's got school-age children then there's very little acceptance around that in a village setting. In a more urban setting, she would be negotiating with neighbours to do after-school shared sessions, there'd be an after-school club or whatever, there's more options in terms of nurseries as well. … If you're a single parent in the village, then you're a mother more than you are a woman. … That's how you're seen, you're seen as the mother of these children not a woman who might have career aspirations. … I know that's a horrible stereotype and I know it's not always true, I do know that, but I've got some evidence that that is the case, it's very difficult to break out of that model.' (Gatekeeper)

Cost of living and access to services

Bellingham and Hexham are the main service centres for those living in the study area, although services in Bellingham can be expensive and their range is becoming increasingly limited. There is no longer a bank in Bellingham (a bank van visits once a week), and post office services have been reduced, making access to cash difficult. The nearest Accident and Emergency is in Cramlington, 34 miles from Bellingham. There is no longer emergency care in Hexham, although a walk-in clinic and many other departments remain in operation at Hexham Hospital.

'The shops, I can't afford to shop in the Co-op. The benefits that I'm on is the minimum. And I cannot afford to shop in the shops in Bellingham. The prices are extortionate. … If I go to Tesco's [in Hexham], I go to the discounted aisle. Something that's got a one-day shelf life left and they've reduced it. But they've got to reduce it down quite a bit, and I'll bring it home and freeze it. … So, I've got meat in the freezer. It's soups and whatnot I mainly get from the foodbank.' (Individual)

Public transport was described as "a very serious issue for older people" (Gatekeeper). Many older people no longer have access to private transport and/or prefer not to drive in the dark or bad weather. The nearest main bus stop for most is in Bellingham, although there is one return bus per day from Kielder to Bellingham. However, the long waits between buses make it unappealing to use. One interviewee described the scenario wherein 'you go down to Hexham for your day out and you sit in a café and have your cup of tea but then there's a long time to wait for the next bus and maybe you're not going to spend any more money than having that one cup of tea and you can't sit in that café for two hours, so it is very restrictive.' (Gatekeeper)

Children and teenagers often travel long distances to school, and it is not always possible for those children who live in the more northerly parts of the area to take part in after-school clubs/activities due to the timing of school transport. It was also noted that, across rural Northumberland, many women may not have access to a private vehicle, as one interviewee explained:

'hardly any of the women have got a vehicle – even when there's one in their household they're quite often not encouraged to drive … so the time it takes to go by public transport and the cost is immense. From Alnwick to Amble return is £7, which is an hour's pay and plus you've got to do a lot of hanging around, so transport is a massive issue.' (Gatekeeper)

As in our other case studies, we heard several reasons why people get into debt and how the rural context and cost of living presents unique challenges. A significant issue, as mentioned earlier, is that many of the county's population experience fuel poverty. Across Northumberland, 27 per cent of

Bellingham is the main town in the North Tyne valley

households were classed as fuel poor in 2018–19 (fuel costs accounting for more than 10 per cent of household income). This is a lot higher than the national average of 11 per cent across England (Northumberland County Council, 2020). About 350 households are off the electricity grid – in neighbouring north Northumberland there about 100 members of one constituency who are in this situation and they rely on diesel generators in some cases:

> 'some of those people want to be off grid, of course, there are people who want the good life type thing, but the vast majority of that 352 we think don't – they can't get out of it, they're stuck.' (Gatekeeper)

Many properties in the North Tyne valley do not have access to mains gas, with houses often not well insulated and exposed to bad weather. Lots of households use coal for their heating, with a local merchant providing an affordable supply.

> 'there's a lot of people who are sort of asset rich and kind of fuel poor … We speak to people quite often that are living within farm buildings within Northumberland and very beautiful properties, places you would probably like to go to for a week's holiday in the summer or something like that but actually, the reality of living there is that the property might be worth a lot of money but then in order to maintain that property, people just can't afford to do that.' (Gatekeeper)

We heard how some households run out of oil early in the winter and cannot afford to buy more. One interviewee described how

> 'quite a few people that we've worked with … have basically ran out of oil and haven't had the funds to be able to buy more oil so they're left kind of over the winter … in a no heat situation.' (Gatekeeper)

This may derive from poor household budgeting and/or the administrative difficulties associated with arranging local bulk-buying clubs, which can make the oil cheaper.

> 'Fuel poverty and the fact that most of those rural properties do not have mains gas at all, they're all relying on electric or oil, which are much more expensive than gas to heat and oil particularly, they are … unless they can pull networks of fuel buying together, which has been trialled, a chap in Falstone wanted to try it, he wanted a garage at Kielder to do it but we realised very, very early on, what he wanted to do was the administration was mammoth and sorting it out was huge and it

would have just cost too much administratively for them to do it. But the only way you can get past the fuel poverty is to have good buy-in schemes and somebody has to do that and organise it and everybody has to buy into the fact that they're going to get a fuel delivery on the second Tuesday of the month or whatever it may be and that they may need to buy three to six months in advance.' (Gatekeeper)

The CAB also advocates on behalf of individuals struggling to pay their energy bills, which has been a very welcome service for some.

'I've been having problems with my electric, something wrong with my electric. I was on the smart meter and it was just escalating out of control. So, I phoned the electric company, I phoned Citizens Advice to see if I could get any help with it. And she put us in touch with somebody at NPower actually who's been really helpful. And then she [at the CAB] said, "Have you tried claiming with this?" I got the same girl that used to see me all the time. She says, "Phone them up and get a claim in". She says, "They can only refuse it". She said, "We'll take it to mandatory reconsideration".' (Individual)

The high up-front costs of purchasing fuel can also present a challenge. Minimum orders of oil (of around 500 litres) "means that that's somewhere in the region of £300 for somebody to order that. And if they haven't been budgeting well, because they've maybe been struggling with other costs, then it's not something that they can pull out their back pocket for the next oil delivery" (Gatekeeper). Although some national support initiatives exist, one organisation we spoke to suggested that people from vulnerable groups who need support are often ineligible for these schemes due to low household income. Similarly, energy advice programmes which exist in cities elsewhere have not reached rural Northumberland because people are too dispersed to make door-to-door advice sessions financially viable.

In this context, budgeting can be challenging, with emergency fuel top-ups, for example, often linked to the wait for the first UC payment to come through. There was a general perception that people have "less to survive" on than they would have had in the past. As one interviewee explained:

'In the past, we could fairly confidently say, if you missed your rent, probably you're missing out on benefits, if you're over-indebted and that money's going to a lot of other places. So, in terms of benefits entitlements, there's always a lot of missing benefit. … You know, extra income-related benefits that people had, and missing premiums. However, over the last few years, obviously, potential for those income rationalisation exercises have diminished and it becomes more about how to survive with very little and budgeting and making emergency budgeting.' (Gatekeeper)

Organisations have responded by providing budgeting advice and support to increase savings. For example, fuel advisers at the CAB offer a service related to fuel debt, through which they 'do interventions, or casework, with priority debt like rent, council tax, fuel debt and so on. We'll sit somebody down, either on the phone or together with them, do a full budget; we'll do a full income and expenditure and a better-off calculation' (Gatekeeper). Fuel advisers also look specifically at fuel debt and make sure that individuals are in receipt of the benefit(s) to which they're entitled, referring people to the debt team as required.

'If an unexpected bill came in, I would have to phone them up and explain to them that I really, really can't afford it, what would be the minimal payment you would accept and I would try and keep up with future payments. It's hard. It's hard really having to budget because that £74 is electric, water, I put £10 worth of fuel in the car, ten- or 15-pounds worth of fuel in the car when I leave Hexham … and then I use it to travel in and out of Hexham. But then I pay bills. My priority ones come first. I've got electric, I need water.' (Individual)

Improving people's understanding of managing their energy usage and costs was seen as a potentially important intervention, as well as raising awareness of the benefits of behaviour change that can lead to increased energy efficiency. Northumberland County Council, the National Park and CAN have formed an off-grid task force to try to get houses without mains electricity on to the national grid. The Northumberland Log Bank, currently operating in Wooler and staffed by volunteers, provides free logs to those who need them. There are plans for a second log bank in Haydon Bridge, which will serve Bellingham and surroundings in due course.

We also learned about an affordable loan scheme offered by a community bank, which highlights the extent of in-work poverty because of low pay and volatile incomes. The most common reason for needing a loan was to buy household appliances. Many people have unaffordable car loans, but owning a car is seen as vital for work in the rural context. The community bank has introduced several initiatives to encourage saving and to change people's mindsets about savings/budgeting.

'What we're trying to do is actually [ask] how much is an affordable loan, with a responsible loan – how much can you afford? That can then be worked out individually that you can actually afford to pay this back but some of that is going to be put in a savings account. People's whole aspirations change when they see that savings pot increase. That phenomenon was part of the payroll deduction. … Actually, some of them said I don't need to take a loan anymore because I'm actually in the habit of saving. It's sort of thinking, okay, we're helping people to start saving but at the end of the day people were starting out who needed access to that loan, but they were actually working.' (Gatekeeper)

Most over-60s in the area were thought to have occupational or state pensions, although this was less likely for members of the farming community who are more likely to work until very late in life. One individual also explained how having savings can be a "double-edged sword" in a time of crisis:

'In relation to foot and mouth in 2001 … my work dried up. And I actually recall going to Citizens Advice Bureau and asking for their help, because I was really struggling, and they looked into my savings and said, "Well, unfortunately …" because I think I had about £8,000 or £9,000 saved up. And unfortunately, because I had that much money, I was only going to get something like £5 a week or something. But, you know, I can't say exactly. It was something pathetic like that. And my savings went down rapidly, because I had to pay rent and that, you know. You have to eat. … So, when you get really bad circumstances like that, it's a bit tough, like. Yeah. … It's just it wasn't worth … in my mind, it wasn't worth the hassle for the … and it truly was going to be a pittance that I was going to receive.' (Individual)

Narratives of place and belonging

At the start of this chapter, we referred to the "history of loss" that one of our interviewees used to describe the events that had happened in the last century in the North Tyne valley. Recounting a "series of blows" that the area and its people have experienced, this detailed account of the area's history perhaps provides some insight into the reasons behind some of the challenges outlined so far in relation to people's **financial wellbeing**, and wellbeing in general. Speculating that all the losses (the railway, changes due to Kielder dam, impacts of foot and mouth disease, the loss of the agricultural mart and other services) are "still very fresh actually in the local people's psyche", this interviewee "wouldn't be at all surprised if mental health, depression and so on comes from a sense of this is a community that's just lost everything", even though things are "on the up now" (Gatekeeper). Another interviewee described the loss of services from Bellingham in particular:

'But having those things [the auction mart and so on] I think brought people from the wider area into Bellingham and Bellingham was a meeting place. When we moved in there were things … there was still a draper's shop. There was a little delicatessen, a baker's still survived. There were two butchers' shops – we've now got one. There's a chemist there. Most of the essential local sales is the post office. And there were two banks. Now the banks have gone. The draper shop's gone. The delicatessen's gone. But the Co-op's expanded. So, Bellingham's still a bit of a meeting place because that's where we've got to go for a pint of milk.' (Individual)

When talking about changes that have occurred in the study area, another interviewee alerted us to a housing needs survey conducted in 1968, which showed a lack of any real change in the issues that people experience in the area.

'it [the housing needs survey report] was in Kielder and Bellingham and various other areas. When you read the issues that it was talking about, you could have changed the year from 68 to 2020 because it was all about young people, access to employment, services declining, no public transport, the need for affordable housing. It was all the same stuff and in terms of vulnerabilities and stuff, none of that hasn't changed or changed much in decades, and that was 50 years ago.' (Gatekeeper)

The loss of younger people from the area (and across rural Northumberland more generally) was also noted, apart from those young farmers who would inherit the family farm.

'With the younger generation, on the whole they're not up in the more remote parts. The younger local people may have had to move away because they couldn't get housing here or indeed employment or to go away to college. They certainly can't afford to buy anywhere and probably can't really rent anywhere nearby.' (Gatekeeper)

The loss of people from the area's communities because "we can't afford to offer the jobs down here" (Gatekeeper) was described as having knock-on effects in terms of the impact on community spirit. With "fewer people, and those people are ageing, the community just starts to weaken" (Gatekeeper). There are "a lot, lot more incomer-type people than your true locals, who worked on the land or had connections with it" and this individual "would quite like to see it hold on to its rural roots, and it seems like they are getting uprooted a bit" (Individual).

Other research within an oral history project has also explored the strong opinions among Kielder residents regarding the 250,000 tourists who visit the village for recreation each year (Skelton, 2014). Those in the community who had lived there for a long time strongly agreed that community spirit had declined, even though those newer residents appreciated stronger social ties in the village as compared with the weaker social bonds in the places they had moved from.

> 'what you do lose, to some extent, in the very isolated villages like Kielder – the community … the ones that live there still know each other, but you get … just the community feel. The activities, the living in each other's kitchens sort of thing, it starts to break down.' (Gatekeeper)

Despite this somewhat negative narrative, it was clear that many of the interviewees had a strong attachment to the valley and were very happy to live there. As one individual explained:

> 'when you're in here, in the environment, you forget how busy other places are until people remind you. … We're actually saying, "No, let's look at what we've got". So you know the grass is always greener over the side? We've sat there and said, "Right, no this definitely might look greener but we're quite happy with this". Then we've built on our strengths rather than looked at the negatives and found them very positive.' (Individual)

The North Tyne joins the South Tyne at Watersmeet just north of Hexham

However, access to housing presented a challenge for some, affecting their sense of belonging and distorting their place narratives. Our interviewees made a distinction between the housing markets in Bellingham, Tarset and Greystead and Kielder in terms of affordability and tenure. House prices are higher in Tarset and Greystead than in Bellingham and Kielder, where people are moving in due to the lower prices.

> 'There is an availability of houses that are reasonably cheap. … they are bought within Bellingham and in the external areas to Bellingham. But there's a funny range of house prices, actually. There are the cheaper houses. The place it's very difficult to find a house is probably around the £300,000 mark. They are difficult to get your hands on. There are quite a few over that, and quite a few at around £200,000 or £220,000, and there are certainly properties in the lower hundreds – £110,000, sort of thing – down in Bellingham.' (Gatekeeper)

> 'They are dirt cheap [in Kielder]. … it's a big sturdy old forestry-built, government-built house, average sized, quite big with a lot of land at the back. … So, somebody who has just sold a three-bedroom house with oil for a dirt-cheap price of £85,000. So that's the affordability.' (Individual)

Access to housing/choice of housing is very difficult for some people and a couple of our interviewees felt that this needs to be looked at more. We heard that there are likely to be several young people who are/have been unable to access suitable accommodation in the study area, and that, across rural Northumberland, incomes do not "match" property prices and getting a mortgage can be difficult.

> 'We did a little count-up and I think there were from Kielder to Bellingham and the sort of rural area I think one day we just sat in a pub and counted 20 young people who would like to stay near here but can't. And that's where the affordable housing or the housing with extra garaging or extra office space attached to the housing is more valuable. Because, you know, a freelance shepherd needs a quad bike and a trailer and various other bits of equipment. Mobile clipping pen and this kind of thing all of which has got to be kept. … So, that person's got particular needs for accommodation that somebody at that level of employment the city doesn't have.' (Gatekeeper)

> 'If you look at that rural element … it is astonishing that we're going to have an affordable housing crisis because the property values will go up, the wages are low, the wages are at times self-employed low

because the poor have got more than one job and therefore they're relying on being self-employed and just invoicing everybody and that means getting a mortgage is almost impossible.' (Gatekeeper)

Private rented accommodation remains an important part of the local housing market, albeit with some challenges noted in relation to housing quality/tied housing and the availability of this type of accommodation alongside increasing numbers of second homes and holiday cottages.

'I lived in an ideal little cottage. It was a farm cottage. And the rent … it wasn't worth any more rent than what I paid for it, but the rent was cheap. … I think it went to £120 a month. But anyhow, if I'd wanted to upgrade into a place that actually had a thing … in that cottage, there was a back boiler to the fire, but there was no immersion heater, there was no central heating, there was no nothing. It rained in, it was damp, it was cold, it was everything else, but it was home and there was nothing wrong with it, really. But if I'd wanted to upgrade, I would have really struggled to pay the rent. I mean, rents nowadays are £400 or £500 a month at the minimum, to my knowledge. And that would have been a huge, huge problem, to find that, and the council tax to pay on top of your earnings. It would have been very difficult at times. There weren't as many houses available, in all fairness. There weren't many available, and more and more of them have gone into weekend cottages, holiday cottages, lets … There will be less available for normal people on average earnings to remain in the area, I would think.' (Individual)

Social housing exists in Bellingham and Kielder but we heard some differences in opinion about the need for/suitability of this type of housing in the study area. There appears to be a general lack of understanding of the need for this type of accommodation in the Bellingham area. One interviewee questioned the demand for this type of accommodation:

'Certainly the majority is in Bellingham. There are maybe five so-called "affordable houses" in Kielder. … But again, you get this dichotomy between housing associations and planning authorities insisting that if you build this, you have to build two affordable houses with it. They didn't fit. They didn't have transport. There's no public transport, so you have to work on the principle that, "Well, why would you put somebody in affordable housing when they can't get out of the village?" So, if you haven't got a local job and there's no transport, then it's pointless having the affordable housing if there is nobody wanting to use it. And they are talking about putting in more affordable housing in Bellingham, but I'm just not convinced that there is the demand for it.' (Gatekeeper)

Another interviewee explained that there is a need for this type of housing but that the council's county-wide allocation policy poses challenges for local people who require access to this type of accommodation.

> 'it's people who can't afford, who are on lower incomes … they can't afford to buy property, the rented property is prioritised for those who are in highest need. … rural residents, they're not homeless, they're not threatened with homelessness, they are living somewhere, they do have some income, they're band three in Northumberland, not one or two, so they don't even get a look in. … the council told me that the local lettings policy trumps the allocation policy, which it does, but only in the categories. So, if you've got a category one in the local area, they'll trump a category one from outside. But if you've got a category two in the local area, they won't trump a category one from outside. Often there aren't category ones from outside that want to move there, so you can still get there but it puts off the community from wanting those houses if they think they're not going to be provided for the need they've identified.' (Gatekeeper)

The allocation policy also presents challenges for the relationships between tenants and local people, due to people with financial and other difficulties being moved into remote rural villages from elsewhere. This appears to be the case in Kielder, where "people are moving into those places who are actually in much more straightened financial circumstances" and

> 'what we were finding was people were turning up with no job, with no car, just basically on benefits, weren't very pleasant people, unable to get a job because they had no vehicle to travel, didn't want to work and spent most of their time in the pub or whatever. … that regrettably bring[s] a very negative town attitude in and with hassle and trouble and arguing between neighbours and things.' (Individual)

Attempts are being made by one voluntary and community sector organisation to work with the council to develop an alternative local lettings policy, which would help to address these types of issues. The focus is on

> 'trying to come up with a model local lettings policy that they'd be happy with in those sort of circumstances, so as a sort of exception … it's not going to have a big impact on the numbers they're going to be able to house across the county, but it would make a massive

difference to those small communities, so that's the slant we're trying to adopt.' (Gatekeeper)

New builds and renovation of vacant properties were not normally allowed in the National Park for many years, unless for holiday accommodation, but this policy changed with the adoption of a new Local Plan in July 2020, shortly after we finished our interviews (The Courant, 2020). This now permits the development of affordable housing on exception sites, along with a limited amount of other new housing where necessary to the National Park's purposes (Northumberland National Park, 2020). This may be particularly relevant in Greenhaugh, where there are no unoccupied properties in the village and a healthy number of people aged under 18. Without the ability to build new affordable properties, those younger people might find it hard to remain in the future. Other research has suggested that the smaller, more remote communities in Northumberland National Park become increasingly unsustainable without new development (Dunn, 2011).

'I mean that is a bit of a problem really because when you get some building that could be some tumble-down place that could be renovated to make it occupiable, as far as I can tell they only give planning permission for it to become holiday homes, that sort of thing. I don't mean full-time holiday homes. I mean self-catering places. That's not really helping. It may be helping the person who owns the building that's being converted but it's not helping the people who would have liked to have bought it to make it into a little permanent home for themselves.' (Gatekeeper)

Conclusion

In the North Tyne valley various changes have led to a smaller population (particularly among those working in the agricultural/forestry sector) and knock-on impacts on available services and leisure activities. The loss of the railway and the Bellingham agricultural mart, the construction of the Kielder dam and the impacts of foot and mouth disease were poignant chapters in the area's history that were thought to be still 'fresh' in local people's minds. The sense of loss was counteracted by people's generally strong attachment to where they live, although this clearly varied according to individual circumstances. There is very strong community spirit in the area, both among long-term residents and among newer residents in the creative sector who have moved into Tarset. In Greenhaugh there are quite a few young families who have been able to remain in available accommodation, despite outmigration being the norm for so many. The sense of community is strong and, for those experiencing financial hardship, that is "the important bit, because you'll get through if you have that spirit" (Individual).

In contrast to Harris, local governance has become more distant in successive local government reorganisations in Northumberland since the 1970s. Despite these challenges, people in the study area clearly have a "do it for themselves" attitude if they want something to happen. This rural outlook was described as different to the attitudes of people in urban Northumberland and perhaps also to those in ex-industrial east Northumberland. There are many examples of positive and enterprising work of voluntary and community organisations in the area, leading to the development of an informal network of community volunteers who themselves bring another layer of connectedness to the support available to rural people experiencing financial hardship.

The area clearly suffers from poor reach by many of the common sources of support, due to its remoteness from larger service centres. Most of the initiatives/formal advice and support points are in Hexham and voluntary and community organisations struggle to deploy outreach services in the study area due to low uptake. Normal channels of communication were not thought to work well for "spreading the word" about available support and advice. Posters in libraries are not generally seen and poor internet access remains a challenge for some. Organisations that are not based in the study area are taking steps to work more closely with "community gems" or "connectors" – people in the smaller villages and towns who informally bring people together and who can help get information to the community. The county council is embracing a place-based approach to understand the nuanced needs in different settlements in the county and making connections between centralised organisations and local connectors is thought to be important. However, these people need support to ensure they offer correct, holistic advice to individuals needing help with UC claims, for example.

Narratives of rural life and place in the study area also intensify the effects of some other social characteristics, particularly age and gender. It remains a challenge across rural Northumberland to reach those who are "proud" and "wouldn't take support". A common theme in the interviews, a reluctance to seek and accept help and support was particularly common among older people. This privacy and self-reliance may be linked with a 'complicit acceptance' of the challenges associated with rural life, such as the 'rural premium' that is thought to be a factor in the lives of the 20 per cent of those who use the foodbank in Hexham and are classed as living in 'perpetual poverty' (ERS Research & Consultancy, 2021). Nonetheless, some voluntary and community organisations have successfully signposted individuals to support, using sensitive language in ways which might 'allow' individuals to accept help. The impacts of public sector job cuts on women's career routes, the disparities in income between men and women and social stereotypes associated with lone mothers also present several gender-related challenges. However, rural women appear to be doing well in micro-enterprise, even though skills training is hard to access, childcare options are limited, they may lack confidence and financial support is hard to access when businesses are not growth-oriented. We reflect on ways to support rural women entering self-employment or seeking to develop their skills in Chapter 8.

Rural poverty in a pandemic: experiences of COVID-19

Our research began in September 2019, before the start of the COVID-19 pandemic, and continued until July 2020, four months into the first lockdown period in the UK. This gave us a unique insight into poverty and social exclusion in rural Britain both immediately before and during the pandemic. Using the evidence from our fieldwork and other research published after the end of our data collection period, this chapter explores the impacts of COVID-19 on rural people experiencing financial hardship.

The lived experiences presented in the three previous chapters highlight the many ways, both positive and negative, that living in a rural area affects local opportunity structures. The cost of living in rural, remote and island areas is substantially higher than in towns and cities, partly because of distance to services, but also because of the costs of heating homes which are often off grid and less well insulated. Access to well-paid work and secure, affordable housing may be more difficult in rural areas without an income from commuting or telecommuting. Access to public services is also likely to present challenges and people eligible for welfare benefits face barriers of distant sources of advice and help, and centralisation of welfare support, inaccessible assessment centres and perhaps social stigma. Private, public and third-sector organisations all face difficulties in reaching into rural areas to offer their support and, while digitalisation may help them to reach some people, this can exclude others without good connectivity or access to devices. These barriers all relate to distance, mobility and access and may be more severe in remote and island areas, like in Harris and the North Tyne valley.

Experiences during the COVID-19 pandemic have brought all these rural vulnerabilities into sharp relief. The Organisation for Economic Co-operation and Development (OECD) deemed rural areas to be particularly susceptible to the negative impacts associated with the COVID-19 pandemic and lockdowns because they generally have: a large share of population at higher risk of severe illness (ageing populations); less diversified economies; a high share of workers in essential jobs (for example, agriculture, food processing) – coupled with limited capacity to do these jobs from home; lower incomes and lower savings; health centres with a lack of specialist services (and long distances to hospitals/COVID-19 testing centres); and a large digital divide (both in terms of access to the internet and connection speeds, as well as fewer people with adequate devices/skills) (OECD, 2020). Rural economies in Britain exhibit some of the features identified by the OECD, including a higher proportion of people working in 'at risk' sectors – those sectors that are impacted by the restriction of movement during a pandemic, such as childcare, restaurants or accommodation services.

In a survey of just over 3,000 rural residents across Scotland in autumn 2020, 24 per cent of respondents were worried about their job security (Generation Scotland, 2021). In the Highlands and Islands of Scotland, where the strong reliance on tourism and hospitality makes the region more susceptible to restricted movement, unemployment increased at a faster rate (118 per cent) than the Scottish average (85 per cent) between March and July 2020 (Highlands and Islands Enterprise, 2020). A survey that received responses from 1,200 business owners and the self-employed in the Highland region found that 54 per cent of the respondents' businesses were closed, 35 per cent were struggling to stay afloat and a further 33 per cent had experienced a fall in sales and profits (Fisher, 2021). Almost half of the respondents were concerned about their ability to survive for the next few months. This percentage of Highland businesses that were closed compared with a much lower figure of 27 per cent in Glasgow and Edinburgh combined. Youth unemployment also continued to rise in the region (from 3.8 per cent to 9.9 per cent in the same period), which highlights significant barriers for young people currently wishing to enter the labour market. This 'pushing' of young people out of the workforce is seen as a global issue that will need national and local responses (International Labour Organisation, 2020).

During the pandemic, small to medium enterprises (SMEs) have been at greater financial risk than larger private organisations/public bodies, with localised, service-based start-ups and micro firms most affected. In Scotland SMEs account for a greater share of private sector employment in rural areas when compared with the Scottish average of 50.6 per cent (for example, in the Highlands and Islands SMEs account for 66.9 per cent of private sector employment; Highlands and Islands Enterprise, 2020). In England 2.6 million people were employed in registered rural SMEs, representing 71 per cent of all those employed by registered rural enterprises, compared with 41 per cent of those employed in registered urban enterprises (DEFRA, 2020).

Self-employment is also more prevalent in rural areas. In rural England 21 per cent of those employed or self-employed already worked from home prior to the pandemic, compared with 13 per cent in urban areas (DEFRA, 2020). It took a longer time for the UK government to frame a workable response to the impact of the pandemic on earnings for the self-employed, when compared with other types of employment. Many self-employed workers were not eligible for government financial support during the first national lockdown that began in March 2020 and self-employed people in the tourism and hospitality sector (especially females) were particularly impacted.[1] This echoes the findings of research by the Standard Life Foundation (SLF) that estimates that 3.8 million workers were unprotected by the financial support schemes and that these were proportionately more numerous in rural areas and towns than in cities (Kempson and Evans, 2020; note that the difference between rural areas and cities was not statistically significant). Among the reasons for people being excluded from the UK Coronavirus Job Retention Scheme (CJRS) were job loss, reduced hours or a recently changed job and (for exclusion from the Self-Employed Income Support Scheme, SEISS) being newly self-employed or deriving less than half their income from self-employment. Many of those excluded from these schemes turned to Universal Credit (UC) (Brewer and Handscomb, 2020), which has its own challenges in rural areas as we have seen.

These issues are amplified by rurality and remoteness, particularly as bank finance is less accessible and public services tend to be centralised. Business growth has been difficult for some women entrepreneurs during the pandemic, especially as it was more difficult in rural areas to find the right people with the right capabilities to work in specific areas such as technology and software development (Arshed, 2021). Digital connectivity also remains a pressing challenge. In a survey of Scotland's rural residents during autumn 2020, 19 per cent of participants described their current broadband connection as 'poor' or 'very poor' (Generation Scotland, 2021). Impacts are also compounded by the relative vulnerability of rural regions to the exit of the UK from the European Union (EU-Exit/Brexit). For example, five of six Scottish local authority areas deemed 'most vulnerable' to EU-Exit are in the Highlands and Islands, due to dependence on migrant workers, EU financial support and having a more fragile population (Scottish Government, 2019). In England the challenges of maintaining and delivering services in rural areas had already been heightened by cuts to English local authority budgets over the last decade, even before any impacts of EU-Exit are felt. The National Audit Office (NAO) found this reduction of around a third in councils' spending power, alongside rising demand for services, had also left councils more vulnerable to the impacts of the pandemic. The NAO warns of continuing cuts to services in the next few years, including social care, special educational needs, libraries, buses and community centres, as councils struggle to meet the extra costs incurred during the pandemic: 94 per cent of councils expect to have to cut spending in 2022 to meet legal duties to balance their budgets, and several risk insolvency (NAO, 2021). This is likely to lead to further centralisation or loss of services in rural areas.

There have been clear 'winners and losers' in rural areas during the pandemic and lockdowns, with different factors affecting the recovery rates of local services (Table 6.1). Locally based food shops were highly valued by rural residents in England and Scotland, for example, with many businesses

Table 6.1: Main rural impacts of the pandemic in England: winners and losers

	Service types	Impact, relative to the pre-pandemic period
Winners ↑	Online services	Existing growth in use made faster still
	Cashless payments	Existing growth in use made faster still
	Rural food shops/village stores	Increased custom from local residents
	Parks and outdoor spaces	Highly valued by many during this period
	Market town centres	Mixed picture, but many recovering well
	Village or community halls	Recovering, but likely some permanent closures
	Cafés and restaurants	Clients mostly returning, but some permanent closures
	Rural bus services	Passenger numbers nosedived, then slowly recovered
↓	Rural pubs	Further additional permanent closures
Losers	Theatres, cinemas and music venues	Initial closure and some audiences hesitant to return

Source: Rural England CIC (2022)

adapting their services to introduce home deliveries (Currie et al, 2021; Rural England CIC, 2022). A survey of rural residents in England revealed that rural communities have made much greater use of their local stores and farm shops since March 2020, with some smaller towns proving more resilient than might have been expected considering reduced high street footfall in the years before the pandemic (Rural England CIC, 2022). The temporary closure of most community or village halls has had significant impacts across rural communities (and in our case studies, as will be noted later), although reduced running costs and the availability of new support grants means that the financial position of many of these halls remains strong as restrictions are eased (Rural England CIC, 2022). It remains to be seen how many groups will continue to use local halls for their activities and how quickly they will return, which may have impacts on income streams in the future.

Despite these challenges, past crises have demonstrated the resilience and adaptability of rural economies and communities – an example is the rural shutdown during the foot and mouth disease outbreak in the UK in 2001 (Phillipson et al, 2020). In 2020–21 the response of rural communities has once again been notable in terms of people working together at the local level to support residents and businesses (Ross, 2020). Strategic partnerships and responsive service delivery have also contributed to effective community responses (Currie et al, 2021). Geography has also played a role, with Scottish islands, like many other islands across the world, escaping the worst health consequences of COVID-19 before the roll-out of vaccination programmes, due to the combination of their geography and their stringent measures (Sindico et al, 2020). Rural regions in Britain have tended to have proportionately fewer cases of COVID-19 than urban areas (at least until the G7 summit in Cornwall in June 2021), for various reasons which have still to be fully understood, perhaps including less mixing on public transport and lower population density. The remainder of this chapter explores how the pandemic was experienced by residents of our three study areas and the organisations working to support them.

Lived experiences in Perthshire, Harris and Northumberland

Like in many rural areas, the economic impacts of the lockdown that began in March 2020 have been felt in all major sectors in the study areas, with the tourism, hospitality and leisure industry hardest hit.

'To give you an idea of what happened during COVID, some of the stats that we heard yesterday [from Perth and Kinross Council (PKC)]: 3,000 calls have been received by welfare benefits and crisis grants, 2,000 food packages are being distributed on a weekly basis through a network of community groups and foodbanks, etc. In terms of unemployment levels during COVID, apparently it's been an unemployment rise of 94 per cent in Perth and Kinross, particularly tourism and service hospitality-type industries. There has been an increase in requests for council tax reductions of 150 per cent. Housing Benefit applications have risen by 67 per cent. Universal Credit applications, according to the call yesterday, there's been a 400 per cent increase in Perth and Kinross. A big increase in rent arrears. The council [also] shared information about social isolation, mental health, relationship breakdowns, domestic violence, digital exclusion.' (Perthshire focus group)

In recent years there has been an increasing reliance on employment and self-employment in tourism and hospitality in all the study areas. This has made many people particularly susceptible to the impacts of business closures in this sector, with job losses expected to have a disproportionate local impact. Unfortunately, we heard that lots of people in the North Tyne valley were not eligible for furlough and were therefore laid off from their work in the early stages of the lockdown. This issue was linked particularly to the nature of rural employment and people having several jobs.

'So, at the moment [in April 2020], we've got a lot of customers who have just been, kind of, cut off, or let loose, by their employers with a vague promise that they'll get some payment at the end of April. So, without there being any furlough letter being given, or change to their employment contract or any, indeed, salary still being paid for the employer to be reimbursed at the end of the period.' (North Tyne valley gatekeeper)

One of our respondents described the PKC Business Barometer Survey of May 2020 which showed how "the impact of the pandemic was felt in all major sectors and localities in Perth and Kinross, with 80 per cent of businesses reporting a loss of income due to the crisis" (Perthshire gatekeeper). Again, it was clear that the tourism, hospitality and leisure sector was hardest hit, with 86 per cent reporting loss of income and 26 per cent planning redundancies: this sector had "lower confidence in their future trading position at ten points less than the survey average" (Perthshire gatekeeper).

'[T]he economic impact of COVID is still yet to be fully appreciated. Perth and Kinross has one of the highest proportional universal claimant rates in the whole of Scotland. Historically, unemployment has been contained to the city centre; however, during COVID, the biggest rise seems to have been in rural and Highland Perthshire because of tourism and hospitality industries in particular. The Scottish furlough rate is estimated to be sitting at around 25 per cent and cities tend to be lower. However, the Perth and Kinross workforce apparently is at 30.8 per cent which is almost a third of all of Perth and Kinross' workforce currently on furlough at the moment.' (Perthshire focus group)

Reliance on tourism and hospitality is also particularly strong in Harris and the North Tyne valley, where there have been many job losses, and the full economic impact is yet to be appreciated. In Harris current reliance on the sector was estimated by participants at between 50 and 85 per cent. Many staff in tourism, in all three study areas, have insecure, casual or seasonal work, with zero-hour contracts common in tourism, hospitality and retail, such that they received little help from furlough pay during the pandemic.

'I think the seasonality aspect is more about the zero hours. So, they may have been on a zero-hours contract, have seen themselves through the winter on their reduced hours and then expected it to pick up just as the lockdown started so they've just had to then go on to the Universal Credit because the furlough pay would have been next to nothing because they've been on zero hours over the winter. Or perhaps the seasonal work was literally seasonal work, they don't have a contract over the winter but they earn enough over the summer to see themselves through or apply for benefits over the winter and then start their job. That tends to be, in my experience of the people that I've worked with anyway over the years.' (Perthshire focus group)

In Blairgowrie this will be a challenge for members of the large Eastern European workforce based there. Some workers (including EU workers) also lost the homes which went with their insecure jobs.

'We've got a large Eastern European workforce that does do a lot of those jobs. What we found … is that a lot of the minority community members that we have seen coming forward have lost their housing and their job because they're very often linked. So that has brought a whole host of different issues along with it. Foodbank use has obviously increased dramatically.' (Perthshire focus group)

In Harris a lot of new businesses were established during the recent boom in tourism before 2020. This is thought to have made the sector even more vulnerable to the recent economic downturn because the levels of debt required for these businesses to start were based on pre-COVID-19 projections of tourism and trading levels. In addition, the Outer Hebrides is ranked highest in the Scottish government's Brexit Vulnerability Index (53 per cent of the region's data zones are in the 20 per cent most vulnerable in Scotland) (Scottish Government, 2019).

Some new business owners in Harris felt they had no choice but to open as soon as it was possible to do so, because of the need to service these debts. At the same time, they feared being the first to open, or to be seen to be encouraging visitors to the island, when the Outer Hebrides had managed up until that point to remain relatively COVID-free. This was particularly the case for hospitality businesses who were perhaps less able to consider outdoor seating options than businesses elsewhere in the country, due to the poor weather and presence of midges (biting insects). This potential exposure to criticism was not unique to the Western Isles, with many rural accommodation providers publicly criticised for taking bookings for visits during the national lockdowns and/or when restrictions were eased and holidaymakers flocked to rural areas for 'staycations' (Maclaren and Philip, 2021).

'So, you've got all these new businesses which have emerged from the distillery and from the marina in Tarbert and Scalpay, new businesses started and were very, very successful but with a lot of money being owed and that hasn't gone away. … I do know that they are struggling, particularly businesses which were set up this year or in the last couple of years with no real accounts to show. There have been a lot of businesses set up in the last couple of years. So, it's a huge worry. A lot of companies are actually concerned about being the first ones to open and the backlash which may ensue.' (Harris focus group)

Animating COVID rules for rural Northumberland: poster in village windows

Source: Poster design from indieretail.uk and thebehavioursagency.com

These employment and business impacts were substantially mitigated in the study areas by the state, notably through the CJRS, the SEISS and through a temporary uplift of £20 per week for those claiming UC and working tax credits. However, many people in the study areas did not benefit from these measures in the first national lockdown, including seasonal, casual and freelance workers and many self-employed, who for one reason or another did not qualify for the CJRS or SEISS.

'I've just spoken to a gentleman who runs his garage out of a farm and so he's had nothing. He didn't get anything because farms aren't eligible. He fell through the gaps. So that's one rural environment problem.' (North Tyne valley individual)

Although some people with multiple jobs are self-employed, they were also missing any form of financial help at the time if their business was not long-established.

'a lot of people are not eligible for any furlough. They're laid off. Other people who actually are eligible to be furloughed but still life is very, very difficult. … those people, or these multiple employment people, they're technically self-employed most of them and they're not eligible for any kind of help.' (North Tyne valley gatekeeper)

This latter issue was particularly linked in the study areas to the nature of rural employment. In all three study areas, there is a tendency for people to need several jobs and/or work casually. In Blairgowrie and the North Tyne valley, this was often as casual or 'loose' farmers in the agricultural sector, whereas in Harris this was more likely to be in tourism or fishing.

'Currently they have no income. So, we've got people actually currently in that situation who would actually have been rushing around and doing a lot of work. A lot of those people with multiple jobs have actually no income at the moment. People who do gardening for other people, people who clean houses obviously can't go into the houses. As I mentioned mobile hairdressers before, lots of self-employed people, tiny, tiny businesses which are quite often people start up something and do it for a short time and then go on and do something else so they wouldn't even be eligible for any of the things that the government are setting out if you were doing it and did a tax return in 2019 because they might not have been doing it and given a tax return in 2019.' (North Tyne valley gatekeeper)

There was also the issue in the study areas of people who were expecting to start their seasonal job, which never materialised. This was akin to having 'three winters'. The timing of the first lockdown, before the season started and people took up their seasonal employment, meant that many of these people also missed out on CJRS. In Harris many individuals and businesses were said to have fallen through the net because of timing.

Almost a third of the workforce in Perthshire was on furlough in the summer of 2020 (compared with the Scottish average of 25 per cent in the same period). Around 24 per cent of the working-age population in the Outer Hebrides (including Harris) was also furloughed at that time (Comhairle nan Eilean Siar, 2020). Interviewees expressed serious concerns about youth unemployment in the study areas, particularly as this is a new concern – in the past, agencies have been more familiar with the challenge of youth *under*employment (in work but looking for additional hours, an extra job or a job with more hours).

> 'The unemployment rate has gone up by 35 per cent. A lot of that is people coming back to the island for the summer wanting to start their tourism jobs, working in bed and breakfasts, pubs and things like that and, of course, there's nothing there for them. So, there's a lot of young people been added to the employment register.' (Harris focus group)

In Perth and Kinross, there were real concerns raised in the focus groups about young people and a potential doubling of youth unemployment.

> 'Perth as a local authority area has never had an unemployment problem. There is an underemployment problem and a seasonal and insecure employment problem but unemployment isn't something that Perth and Kinross has ever really traditionally had to battle with, so this is a completely new level of learning and intense support that's going to be required.' (Perthshire focus group)

The lockdown also affected the claimant rate quite dramatically in the study areas. The rate more than doubled in Perth and Kinross in the first two months of lockdown in 2020. Before lockdown, the geographical spread of claimants in Perth and Kinross was split quite evenly between Perth City and rural wards. By May 2020, 56 per cent were in rural wards and 44 per cent in urban wards, suggesting that the restrictions related to the pandemic had a proportionately higher impact on the working-age population in that area (PKC, 2020). There was also a significant increase in rent arrears across Perthshire. In the north-east of England, there was also a dramatic rise in the claimant rate, slightly higher than the England average (Figure 6.1).

In the Outer Hebrides applications for UC initially increased threefold due to COVID-19. Although the number of claimants had reduced slightly in December 2020, it remains unclear what impact the extension of the furlough scheme and changes to UC claims will have in the area (Comhairle nan Eilean Siar, 2020). In Harris a lot of people had to apply for UC when they had never encountered the system before. They were said to have been 'shocked' at how low the benefit payments were. It is likely that this will change the perception of many people about the welfare system in general. Participants thought that many people in Harris tried to survive on their savings at the start of the crisis because of the perceived complexity of

Figure 6.1: Benefit claimant rates in North of Tyne, North-East and England (2018–21)

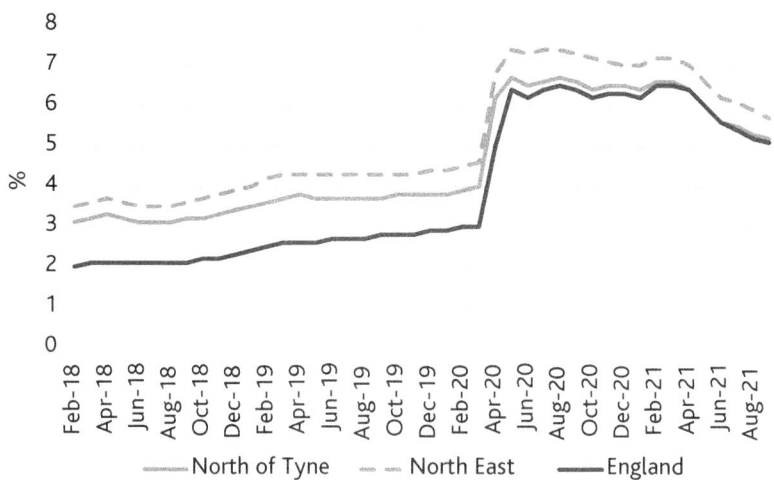

Source: Hamilton (2021)/ONS Official Labour Market Statistics. Contains public sector information licensed under the Open Government Licence v3.0.

the welfare support system and the time and effort needed to navigate it. However, it was likely that more people decided to apply for support as the pandemic continued.

The increased claimant rate placed severe pressure on UK Department of Work and Pensions (DWP) staff and the Citizens Advice Bureau (CAB) and other advisory services working with rural claimants. In Harris demand for advice services from new clients was reported to have increased by about 50 per cent, and representatives of the advice services felt that they were reaching people that they had been unaware of being in need previously. Many agencies across the Outer Hebrides saw an increase in the number of clients with employment, financial and fuel poverty concerns (Comhairle nan Eilean Siar, 2020).

> 'I've looked through our stats and we noticed the most common area is benefit advice. Now, we had a big rush because there's a lot of the fishing industry around there and they were told that when lockdown came their jobs would go completely because the markets in Spain, France and places like that were closing completely. So, I think we'd arranged to close down on the Friday. When the fishing companies closed down on the Wednesday, we had a massive rush of people coming to us one to one because they didn't know how to use the IT to claim and a lot of people there maybe with English not as their first language. So, we had problems there. So, what we did, we actually extended our opening period to cope with this.' (Harris focus group)

Concerns were also raised across the study areas about delays in UC payments, which is not an issue unique to during COVID-19 and has received attention in earlier chapters. However, the pandemic has brought this issue into focus once again.

> 'It really has been the delay in Universal Credit for people has been a big thing around here. I know there's three families on my street who had to move on to UC when this started.' (Perthshire focus group)

People who were already receiving welfare support faced less of a challenge than new claimants who had to learn how to negotiate an unfamiliar and complex system, mostly online and often without access to an internet connection or device. In non-pandemic times, claimants in the study areas were likely to be disadvantaged by the combination of centralisation and digitalisation of the benefits system. This can be because they lack access to broadband, can't afford broadband or because broadband is of poor quality.

> 'Blairgowrie library has got a Wi-Fi connection and outside the council office. If you hang about outside the council office or outside the library, you can access the Wi-Fi signal if you're all registered and you can do the PKC thing and register it altogether but there's nowhere else. You can't just hang about … It's a really big disparity between being in the middle of Blairgowrie and even 500 metres away from that area and having Wi-Fi spots available for people. I don't know if it's better in Perth City. I think it is. I think you can pretty much walk around Perth city centre and access various Wi-Fi spots.' (Perthshire focus group)

They can also lack digital skills, and/or experience literacy or mental health challenges, which presented challenges both before and during the pandemic. For example, when describing the challenges faced by a lady in her 60s in the Blairgowrie study area:

> 'Welfare rights and Perth and Kinross Council are great in terms of giving advice – it's very much an online system. You have to phone a number which the worst thing for this lady is when she phones up she gets a whole list of "choose this number, choose that number" – by the time they've got to number three, she's forgotten what number one is. So, she needs vast amounts of support just to get through that point where she's emailed, she's taken a photo, she's been able to do all this. We were, finally, I have to say, at the point where things were balancing out and then lockdown happened.' (Perthshire focus group)

An additional challenge is the greater difficulty in travelling from rural areas to find face-to-face support. The lockdowns have brought the variation in the quality of broadband into sharper focus, demonstrating inequality among those learning or working from home with poor connections or a complete lack of connection to the internet prohibiting people's ability to apply for state support during the crisis.

'At the moment I'm supporting two women who, in the last two weeks, one was made redundant and the other one has run out of money. … So I've been supporting them getting on to Universal Credit, but in order to do that, neither of them have got internet at home … I've had to meet one of them on a park bench – we had a really bad time of trying to get on to the gov.uk Universal Credit website where you can sign on. So, neither of those two women would have got signed on without that and they're both absolutely bankrupt, they're both penniless.' (North Tyne valley gatekeeper)

One of the exclusively rural impacts of the pandemic arose out of the more limited provision of suitable specialist health facilities in remote communities, particularly the islands. This has brought the precariousness of rural healthcare (especially in remote communities) into sharp focus (Maclaren and Philip, 2021). For residents in Harris, access to intensive care beds is very limited, and for an island group with a much higher proportion of older, and therefore more vulnerable, people this could be catastrophic for residents across the Outer Hebrides. At the time of the fieldwork, the levels of infection in the Western Isles had remained very low because access to and from the mainland via plane and ferry had been severely restricted.

Although local councils moved quickly to work with community groups, the National Health Service and other agencies to respond to emerging needs, support for care at home was often withdrawn or reduced due to the social distancing restrictions, so impacting on many vulnerable people and their unpaid carers. Questions were also raised in one focus group about how well the various support organisations have been able to understand the experiences of those receiving and providing care roles.

'I think the access to services and the voice is maybe being less … or it has not been heard as much from carers or from older people whose services just stopped and nobody told them, apart from, "Oh, sorry, Jesse, it's COVID. We've not got the staff anymore. Your two visits are down to one visit". And nobody told Jesse that there was going to be an alternative or that something else would happen, so Jesse says, "Okay, I understand. You must go and help the people that have got COVID. I'm 93, and I did have three visits, and now I've got one visit. I'll be okay". I think that that has not come through, so I think their vulnerability wasn't pushed forward enough. And I think carers who just thought, "Actually, I can't have anybody in. I'm scared you're going to give my son, my mum, whatever …" and so their voice has not been heard. So, I think that the impact on them, we don't really know how much it is, because maybe we've not heard them as much.' (Perthshire focus group)

Many participants in our research were concerned about the wellbeing and financial vulnerability of unpaid carers. Particularly hard hit were those caring at home for loved ones, whose support was reduced or withdrawn for a long period, leaving many exhausted, mentally and physically.

'[W]e have about 20 families with children with disabilities. And because they weren't given that label of "vulnerable family", their support was withdrawn completely. So, we've got families that have got three disabled children. And most of these children are either at a special school or they have support that is different. … All that went. So, you've got families who are exhausted, because they have been doing this for three months now. They are all tired. And you've got children whose routines are completely lost and their connections, especially children with autism and who have had quite a little programme. So, I think that unpaid carers in some communities … have taken an absolute hammering over COVID, and I am deeply concerned about their wellbeing, their vulnerability and their financial vulnerability too, because some of them may have been able to be paid for that role. … And we also had critical workers' families who were expected to take their children down to Perth. So that's definitely a rural issue.' (Perthshire focus group)

One of the main concerns of government during lockdown was the impact of school closures on children, especially from poorer families, as well as on the ability of parents to go to work or work from home. Children were supposed to be able to access education online while at home, while children of key workers were still permitted to attend school. Rural contexts proved challenging in both respects. Limited access to broadband and/or a device was once again an issue in parts of each study area where slow download speeds were a barrier to accessing home learning materials while schools were closed.

'I think the education thing has been a real shocker for me, as a parent, the lack of preparedness for any of this really, really surprised me. It shows you how much is based on in-class learning before this all started and the low tech that's in schools is a big problem. The Wi-Fi access in [X] is so bad that my kids very rarely get the internet at school and then to expect us to all be able to log on and be part of that. I'm still reeling from it to be honest. I think it shows a big disparity in rural schools and how much resources they have, even compared to Rattray. I hope that from this point on they think far more innovatively and get far more resources into these schools if they are planning to have any long-term impact to change how much has been lost.' (Perthshire focus group)

There were also language issues in families where English is not the first language, with parents in those households not able to access normal learning/language support.

'One of the things we've raised recently is about all the kids that obviously haven't been in school and all the assumptions about, "Oh well, you just log on to Glow on your device and you can keep up with all your schooling". Well that's not working because some kids don't have devices. Some kids don't have access to the internet. Some kids don't have access to a space to sit and do their learning and all that type of support. If English is your second, sometimes your third language and you're needing all that additional support then the inequalities that were there prior to COVID are only going to be exacerbated when things return.' (Perthshire focus group)

During the first lockdown in 2020, face-to-face support was not allowed under the COVID-19 restrictions, which created additional problems in supporting some people in the study areas, for example, those with literacy or mental health problems who need to claim welfare support. Across the study areas, participants noted that lockdown was really challenging for older people and those with mid-range mental health problems who require support but who cannot meet advisers face to face and cannot navigate phone or online services.

'It's just really hard because there is no face-to-face anything at the moment. Some people just really can't do the online telephone thing and that's going to be a struggle, that there's going to be no face-to-face support unless it changes and guidelines change and they decide to put that back in place.' (Perthshire focus group)

In the North Tyne valley one individual described the devastating effects on his mental health due to the closure of the foodbank in Hexham. This meant that he was unable to talk to the staff/other foodbank users in person, although food deliveries continued. The Blairgowrie foodbank was forced to close for five months in 2020, and there was a stark reduction in volunteer help (Kelleher, 2021). In Blairgowrie we learned about an elderly woman who had, at the time of the discussion, spent 16 weeks by herself, except for the weekly visit by a volunteer with a food package. Just as in pre-COVID times, the social aspects of these support services were almost as important as the practical support offered, but further heightened during lockdown, highlighting the very negative impacts of the cessation of community-based activities during the lockdown.

'I've got a lady … it's not about money sometimes. It's not always about money. It's about the fact that this lady has now spent 16 weeks nearly by herself. If it means I send a volunteer around once a week with a little food package so they can have a chat outside the front door and she can make her feel better and not feel lonely, it has the same value. I was trying to explain to my team that that has the same value as giving food … it's £10 a food package. It's not like we're giving her £50 a week but what we're giving her is that human contact. That's what a lot of the older people in Blairgowrie are really suffering from now … because there were so many groups in Blairgowrie. I mean there must be nearly 100 different community groups in Blairgowrie that on a monthly basis, on every day of the week, if you want to do something community-based or go and see a group or do something in Blairgowrie, there is something for you. Of course, that has all gone and that's a big change for our elderly population. We're going to see that coming as it goes on and goes on.' (Perthshire focus group)

During the lockdown, travel restrictions also created additional challenges for rural and island residents in accessing cheaper food. In all study areas, residents felt unable to travel to larger towns to shop at the cheaper supermarkets, particularly as public transport use has been strongly discouraged and timetables reduced.

'Some of these smaller localities, some of these smaller hamlets have got no public transport. If you don't have a car then right now you can't even go to the local shop. Forget the five-mile rule, you can't get anywhere in five miles around here … Right now, during COVID, there is no public transport [anyway].' (Perthshire focus group)

'[During lockdown] people who previously had been managing on benefits were actually having to access food support, maybe not every single week but certainly on a regular basis once or twice a month because shopping locally was much more expensive than going to Aldi in Perth. Of course, in the early days, there was no online shopping slots available for deliveries. … I think that's one of the reasons for the demand on foodbanks and things are slowing down, is that people are beginning to be able to access the larger cheaper shops in town [again].' (Perthshire focus group)

In Harris during the early weeks of the pandemic, the availability of some key food and household items was compromised because of 'panic buying' on the mainland. The local foodbank reported struggles accessing supplies from the mainland during this time, which was also a time of much greater need with demand for food parcels having significantly increased. Part of the increase in demand was because of a high number of people shielding (staying at home with very limited contact with others), but it was also because of an increase in the numbers of people in financial difficulties who were struggling to provide for themselves. Local shops implemented delivery services for people shielding or those who didn't have access to transport, in order to ensure they received regular food. These services are new since the pandemic started.

'Our last order was for about £16,000 that we wanted. It's all tinned, packets and stuff. Of that £16,000, they couldn't supply about £6,000 of what we wanted. So, things like tinned vegetables, coffee, cereals and things like that, they couldn't supply it.' (Harris focus group)

The impact of COVID-19 in terms of fuel poverty and debt is likely to become more apparent over the long term following the winter of 2020/21 and considering ongoing uncertainty over energy prices, business security and employment. Although aggregate household savings across the UK rose sharply during the pandemic for many households, this increase was mainly in high-income households with those in the bottom quintile seeing a net reduction in their savings in 2020/21 (Hamilton, 2021).

Responses in a crisis

Across Harris, East Perthshire and the North Tyne valley, individuals, groups and organisations responded in many ways to the impacts of the pandemic and lockdowns on people experiencing financial hardship.

In relation to support from the public sector, there was general praise in the study areas for the way that DWP staff rose to the challenge of processing the wave of new welfare support claims, reportedly "abandoning the bureaucracy" and suspending many of the difficult elements of the welfare regime such as sanctions and assessments temporarily while adopting a more generous and supportive culture.

'I think [DWP] staff have gone out of their way to go above and beyond a lot on a personal level as much as they can within the parameters that they've been given. I think that's the same for anybody that's in a situation where you're confronted by people who genuinely don't know where their food is going to be coming from in the next couple of days. I think there's a lot of us who probably bend rules and boundaries that are meant to be there, but we can't see them at certain times because that's the best outcome for the people that we work with. I think DWP, just from some of the conversations that we were having yesterday, they seem to have abandoned some of that bureaucracy.' (Perthshire focus group)

Participants in Harris applauded the response of the DWP because a process was introduced quickly for claimants to receive an advance on any payments, and these could be paid back over a longer period of time than normal. Medical reassessments for Personal Independence Payment (PIP) were also postponed, but payments were not stopped or delayed until the reassessments could take place. All medical assessments were completed by telephone. Although this mitigated some of the travel issues associated with assessments in distant locations (outlined in earlier chapters), phone appointments presented specific challenges for some people.

'We had recently an issue, a lady who has mental health problems and quite a lot of anxiety issues as well. She had been told the day before that she will get the assessor to call her at 2:15 pm. Well she doesn't pick up the phone unless she's told that at this time she's going to receive a call. So, somebody from the DWP tried to call her on a few occasions in the morning: she didn't pick up the phone. Then at 1 o'clock she actually picked up the phone and was told that she'd actually missed her opportunity to have the assessment, therefore the assessment has been postponed for another fortnight. I think maybe somebody has tried to do a favour to the lady giving her an earlier appointment but they didn't actually put that there is another factor that might be a problem in this particular situation. People with mental health problems definitely do experience trauma for this telephone conversation. Not that face to face isn't also traumatic [as well as travelling to Dundee or Stirling]. Some of the assessments are really traumatic ... but I think the telephone option being the only option at the moment is an option but not always perfect for an individual with mental health problems.' (Perthshire focus group)

For residents of all three study areas, the CAB switched fully to dealing with enquiries by phone and online rather than face to face, pooling staff nationally to manage the enquiry lines. While this was a helpful adaptation at the national scale, some participants in Harris felt that existing clients were being neglected as demand rose, and that it was harder to deliver the same quality of service without seeing people in person. In relation to Blairgowrie, there were serious concerns about the impact of library and other closures on people's ability to access the internet to complete a welfare benefit claim.

> 'there is a saying in CAB that you don't really know what the true problem is until you've been talking to somebody for half an hour. So, a lot of the tools that we work with regularly have been taken away from us a little bit.' (Harris focus group)

There was informal joint working between health and social care staff during lockdown where each would cover some of the other's roles and responsibilities during home visits without seeking permission from above or worrying about whose budget was paying, with improved efficiency and care. The greater autonomy of frontline staff and freedom from bureaucracy was widely valued.

> 'with COVID, what we've done is we had a two-year plan that we've done in five weeks, which is quite impressive because it needed to be and we've got a centralised number, which we did try and have an infrastructure host – I'm very big on pushing things into the community and supporting and enabling capacity building there.' (North Tyne valley gatekeeper)

There were also innovative, local approaches to social care provision, delivered by community groups. Several examples from elsewhere in rural Scotland, including 'Kirrie Connections', 'Out of the Box' and Voluntary Action Orkney, were presented at the Scottish Rural Parliament in March 2021. One focus group included a discussion about creative forms of care provision in the Perth and Kinross study area:

> 'Care at home does not need to just be a heavily regulated statutory service or a private contractor. Care at home can be delivered by micros. It can be delivered by … well, actually, care at home [has been] delivered by a lot of community groups over the last three months.' (Perthshire focus group)

Voluntary and community organisations were often the first port of call for "kind, compassionate and empathetic support" (Perthshire focus group) and for emergency supplies of food during lockdown in the study areas. For the most part, these organisations responded quickly and effectively to the crisis, helping people access welfare benefits at the outset and then continuing to help those who fell through the cracks of state support. The network of 36 Warm Hubs across Northumberland had to close their doors to members of the community but some continued to offer a food delivery service, telephone advice and emergency phones. For many people, access to support from these organisations worked well enough by phone or email. Apart from providing advice, voluntary and community organisations also arranged food parcels, home deliveries of prescriptions and shopping, and phone calls to vulnerable people, with the help of large numbers of new volunteers, predominantly of working age. There were aspirations among our research participants that this engagement of younger volunteers might continue in the future. However, during the second UK-wide lockdown, imposed in late December 2020 after the completion of our fieldwork, stories about rural volunteering and 'communities rallying round to face the lockdown together' did not seem to hit headlines (Maclaren and Philip, 2021). It remains unclear whether the response was more muted in the second lockdown and whether the increase in volunteer effort and wider community engagement will continue post-pandemic.

'We were astounded by the volunteer response, absolutely astounded. As someone who has been trying to get volunteers onboard … for the last two years of my job, it was a flip around. People went from, "I might volunteer if I had a few hours" to, "I'm volunteering. What do you need me to do? Give me something to do. I want to do it now".' (Perthshire focus group)

'One of the things for us … the issue around volunteering and the age of volunteers, where most volunteers do tend to be older, a lot of volunteers have been dropping off anyway, because older people have had childcare responsibilities or they're still working, so therefore not available to volunteer but they're all now self-isolating largely, so interesting that new volunteers are emerging now under this current crisis. So that might be something that helps going forwards. I probably would have said a vulnerability has been the fact of reliance on older volunteers for delivering whatever services they might be delivering – I suppose it tends to be to older people.' (North Tyne valley gatekeeper)

It is notable that the recourse to foodbanks increased very substantially during the crisis in each of the study areas, despite the challenges of supplying food to people in rural locations. This is not unique to foodbanks, with many local food suppliers/shops seizing the opportunity to deliver food to their local customers.

'Bellingham down the road has been able to keep things moving with their local shops. A gentleman we always have that delivers to us, he's now being run off his feet because he was able to fall through the gaps between Tesco. So, he was able to do that. He got his goods from Newcastle and would bring them in to us. Now, we've always used him but instead of turning up at 10 o'clock in the morning, he's not turning up until four or five in the afternoon because he's so busy. It would be nice if that continued.' (North Tyne valley individual)

The ability of voluntary and community organisations to respond in the study areas depended not only on ingenuity and a growth in the number of volunteers, but also on the financial strength of the organisation and its assets. The community trusts in Harris had to furlough their staff and suspend operations to survive the loss of tourism-related income, and community councils had a much more central role. However, neighbouring trusts in the Outer Hebrides with revenue from community-owned wind farms were able to expand their operations, and indeed coordinate support to their communities (Currie et al, 2021).

Those reliant on revenue from suspended activities in community halls, charity shops or fundraising events suffered significant loss of income, in some cases threatening their financial sustainability, although in Scotland grants such as the Third Sector Resilience Fund, the Community Recovery Fund and Adapt and Thrive provided essential funding to voluntary and community organisations to stay afloat during lockdown.

'I had an email from [a local hospice] a couple of days ago … because their charity shops have all shut, all their fundraising has been shut down, at the end of the first quarter of the year, they'll be £1.1million short. Well, blimey, if you're talking about that sort of level of income you've got to be very well organised and aggressive about how you're pursuing it.' (North Tyne valley gatekeeper)

In all study areas, there was some overlap/confusion in relation to how the work of voluntary and community organisations and the local authority complemented each other. In Harris the role of community trusts was sometimes uncertain, with some statutory agencies not knowing where such trusts 'fitted' in the local governance jigsaw, or their position being variable depending on the individual community trust. While there were very local, democratic organisations able and ready to provide community-based support, these groups were not necessarily always used to their full potential and there were frustrations about the lack of joined-up working between these organisations and the statutory agencies.

'One of the things I think that this whole experience shows clearly is at least an opportunity to start to think about hopefully quite radically changing, or making the case for changing, different sorts of relationships and allocation of resources, particularly within the rural context, to think about how you can feature and work with community trusts as a much more central part of that localised place-making process.' (Harris focus group)

Voluntary and community organisations were also impacted by the closure of community buildings, which was seen in one focus group as a missed opportunity to offer the support that people needed while simultaneously strengthening a joined-up response with public agencies.

'I was disappointed that a lot of community buildings were closed during this. I understand why … it happened, but it would have been so useful to have been able to have access, for example, to school kitchens and community buildings during this period in time. … my disappointment was, I think it was a good opportunity for the council and some of the other partners to be a stronger part of that and therefore, kind of, tighter in together, you know, beyond.' (Perthshire focus group)

The national pattern of communities stepping in to make sure their friends, family and neighbours were looked after during the first lockdown in March 2020 was replicated in the study areas. In Harris interviewees felt that the situation had brought greater community cohesion and support. We were told that the community had mobilised during the pandemic, making sure that people were cared for and there was someone looking out for those who were on their own, or shielding. Some commented that the view that everyone in Harris knows everyone else was not quite true anymore, but that the pandemic had transformed that and had brought much greater community cohesion and support.

'I think the mutual support thing has been probably … the most transformational thing throughout COVID for us. … Whilst there was always that culture there of neighbours helping each other out, I think sometimes that's assumed of the Western Isles and it's not actually always real. I think we've had lots of people moving around. People don't mix in the same way that they did 20 years ago. … we have groups where people know everyone in their street now and they didn't know people at all beforehand. … it's been the most transformational thing … for probably the last hundred years.' (Harris focus group)

In the North Tyne valley many residents are used to surviving and having food stocks at home, in preparation for being snowed in during winter months. There were many examples of community members working hard to support each other, as well as businesses continuing to provide local services and expanding their offering and reach as needed.

'People are pretty sensible up here, but they have also been through a lot. I mean, it's a harder life, if you like, in the North Tyne. It's not unusual to be isolated for two or three weeks, if you get a bad winter. So, in many ways, we were better prepped for it anyway. Everybody that I know would have had at least two months of supplies in the freezer.' (North Tyne valley gatekeeper)

'I was speaking to a friend of mine within the church group up in [village], a couple of days ago, and she … she isn't actually self-isolating, particularly. I mean, she is isolating, but she is not ill or anything. And she said, at the moment, she is having to fight people off to do her shopping. You know, people are getting quite disappointed when she says, "Oh, somebody is already doing it for me".' (North Tyne valley gatekeeper)

The situation also enabled some people to ask for help perhaps more easily than in previous years. It appears that less stigma was attached to claiming UC or even to accessing food parcels and free school meals as the nature of the crisis absolved recipients of blame. It was suggested in all three areas that the impact of the pandemic may be to shift perceptions of stigma in rural areas, making people more open to support of one kind or another in the future. Many people celebrated the desire and willingness of people to offer help, whether through formal or informal volunteering or through everyday acts of kindness, and for many this reinforced a dominant discourse of caring, self-reliant, resilient rural communities.

'If you give the community the say-so to do it, they will do it. In a crisis they'll do it. I don't know it will impact as we come out of lockdown and more people go back to jobs or go back to their normal lives – that's going to be interesting to see how much volunteering drops off again.' (Perthshire focus group)

Families, friends and neighbours also offered vital support to households who were shielding or otherwise vulnerable. Access to such support may still have varied according to people's social relationships within the community, while those with family members further away may have found it more difficult to benefit from their support during lockdown.

'But I also think the resilience … I'm on a street with five families and for 12 weeks we never saw a kid outside because everybody stuck to it. The kids were brilliant. They were resilient. They found things to do in the house. I know it's been terrible for some families but for some families, honestly, I'm so proud of them. They did so well.' (Perthshire focus group)

Looking forwards

The pandemic and associated lockdown restrictions presented a complex situation in Harris, East Perthshire and the North Tyne valley. As in other rural areas across the country, residents and representatives of organisations offering support were uncertain about what the future brings for families and communities.

'by the end of the year we're anticipating a tsunami, avalanche, tidal wave of needs coming through the door. It is only going to get worse. I know that a couple of the hotels across [location] have already closed and made people redundant. As someone said, a lot of those are migrant workers. A couple of hotels in [location] have already closed as well. So yes, it is going to get worse. It's not going to get any better. It's going to get a lot worse with the need increasing over the coming months.' (Perthshire focus group)

It was thought that the process of recovery could be very slow. The full impact of the pandemic and lockdowns is unlikely to be felt for some time, while the effects work their way through industries associated with the dominant tourism and hospitality sector (for example, local retail outlets, food suppliers, trades and services for tourism accommodation, car hire companies and so on). Although official forecasts expected unemployment to remain at pre-pandemic levels when the CJRS ended in September 2021, there were still 1.1 million workers on furlough at that point (Brewer and McCurdy, 2021). These workers were predominantly in sectors like aviation that are yet to return to pre-pandemic levels of activity and include a lot of older workers who tend to find work after periods of worklessness. It was anticipated by our respondents that the prevalence of specific industries in rural areas is likely to make these places more vulnerable and the process of recovery more uneven, with the situation likely to get worse before it improves. Youth employment was thought to present a specific challenge in rural communities and this, together with increasingly unaffordable rural housing, will continue to lead to youth outmigration from rural areas. However, British tourism boomed from 'staycation' demand in the summer of 2021, before suffering further lost business during the Omicron wave during the winter of 2021–22.

'A lot of people who are now expecting, when the furlough scheme ends, to lose their jobs and most people that I've spoken to expect to lose their jobs, they don't expect to be brought back to work. I think that's where our worry and our planning is going to come in in the next few months is that food needs are going to become higher and access to the foodbank around here and any of the food projects that have just been at the side of the foodbank, they will need an increase in capacity.' (Perthshire focus group)

There remained frustrations among some of our participants about the oft-untapped potential for local authorities to support other organisations working to help those in need, but there are also examples of large-scale efforts to link up community support (for example, via Northumberland Communities Together).[2] There are clearly lessons to be learned for the future, both in terms of clarifying the roles of different organisations/ community groups, building up relationships and having greater clarity about positions and functions.

'One of the things I think that this whole experience shows clearly is at least an opportunity to start to think about hopefully quite radically changing, or making the case for changing, different sorts of relationships and allocation of resources, particularly within the rural context, to think about how you can feature and work with community trusts as a much more central part of that localised place-making process.' (Harris focus group)

In terms of sources of support, for most people the loss of earnings from the labour market has been mitigated by additional support from the state, often accessed with the help of advice and support from voluntary and community organisations and with a kinder and more generous approach from DWP. Many people still fell through the cracks in the support offered by the state, however, including many of those in insecure employment in rural areas, many EU workers and many self-employed and freelancers. Apart from tightening their belts or borrowing, they were helped mainly by voluntary and community organisations, notably bringing food and other necessities from foodbanks, or by neighbours, friends and family. The important role of voluntary and community organisations in supporting rural residents has also been striking throughout this research. This raises questions about the future impacts on rural communities if voluntary and community organisations (as well as local authorities) face financial difficulties.

'I am concerned in terms of the third sector as to what needs to happen next because I think response is one thing but actually, the third sector is in an absolutely shocking situation in terms of how it's going to survive. A lot of organisations and charities are thinking they're going to be bust by this time next year. What happens next is my biggest fear in relation to that.' (Perthshire focus group)

The COVID-19 experience has strongly highlighted the everyday importance of digital exclusion, the continuing loss of services, the fragility of social care provision and the vulnerability of particular social groups including people with poor literacy or poor mental health, and people with precarious employment conditions (especially EU citizens).

'And I think as resources are getting tighter, and I dread to think what's going to happen after all of this has calmed down, if it does, but resources are going to get more and more tight. And services are going to become more and more centralised, leaving those people who are living out in rural areas even more isolated and even more vulnerable.' (North Tyne valley gatekeeper)

On a more positive note, the situation has shown that many people can work from home. Notwithstanding issues related to broadband availability and quality, this opens up the potential for more people to work from home in rural areas, perhaps helping to stem issues associated with outmigration and depopulation. However, such a shift could place greater pressure on the rural housing market if people decide to opt for a more rural, 'safer' lifestyle, unless more affordable housing in rural areas is prioritised and enabled. It could also mean that people from all walks of life, from business to government, could be located in rural areas and still do their 'central' jobs, which would give them a much better understanding of the reality of rural living.

'That has always been the problem. I bet you that hasn't changed in 26 years in that people were on the fringes or always remote from the agencies who are always based in the centre. Well now, thanks to COVID-19, we know you can be anywhere and you can have a proper meeting with your higher professionals so why should the agencies in fact bother with buildings in Stornoway or Inverness or wherever and just have people all over the place and be meeting virtually? If they were embedded in communities then it would make a massive difference, I'm sure.' (Harris focus group)

The experience of telephone/online delivery for advice services has also shown that this mode of delivery can work for many in rural areas, especially for ongoing communication after an initial face-to-face meeting. It has also shown that staff can work remotely. There was optimism among participants that they will be able to offer a more accessible and inclusive service in the future. One of the barriers to doing this pre-COVID was the lack of access to reasonable IT equipment and a reliable internet connection, and these issues have to some extent been addressed as a consequence of the pandemic.

Combined, these learning points are leading to greater confidence about how the advice services can better deliver services in the future. It will be possible to make greater use of telephones and a network of people, or IT facilities, spread across rural and island areas in existing accessible locations, rather than requiring all clients to come to one central office. Nevertheless, some people will still require face-to-face support and it is vital that they are not forgotten or excluded.

Many people in the study areas celebrated the desire and willingness of people to offer help, whether through formal or informal volunteering or through everyday acts of kindness, and for many this reinforced a dominant discourse of caring, self-reliant, resilient rural communities. Several hoped that a legacy of the COVID-19 experience would be a greater recognition by those in authority of hidden rural poverty.

'I think that the hidden rural poverty has now been recognised as an issue by the council because of the numbers seeking food support and the Universal Credit and [that] has led the council to recognise that there is an issue about hidden rural poverty and it's great that that is now being recognised.' (Perthshire focus group)

Conclusion

The national lockdown that began in the UK in March 2020 delivered a huge shock to rural economies and societies, most obviously through the temporary closure of many businesses and the loss of earnings to employees, self-employed and freelance workers. While it was known pre-pandemic that a substantial proportion of rural residents are at risk of poverty and experience financial vulnerability, our research during the pandemic found widespread worry that more rural residents will be at risk of financial hardship and vulnerability in future, as the full impacts of the pandemic play out and sources of support become more constrained. This has been echoed in other research in Scotland that has noted the need to support the hardest-hit people who live in remote and deprived areas, particularly in relation to food and fuel poverty and rising inflation and/or those managing on UC (Bryce et al, 2021). Across Britain, the most vulnerable citizens have been disproportionately affected by the pandemic, widening the gaps between different groups in society (Boyle, 2020).

A larger proportion of people in our three study areas worked in the tourism, hospitality and leisure sector, and in other sectors linked with precarious and often low-paid employment. Moreover, we found evidence of many people, including seasonal, casual and freelance workers and many self-employed, who did not qualify for one reason or another for the government's support schemes, and so were part of the estimated 3.8 million 'excluded' nationally. In short, many rural residents are highly financially vulnerable as the cost of living crisis gathers pace and will be at risk of poverty unless appropriate action is taken.

This evidence of people's experiences during the COVID-19 pandemic reveals again the different local opportunity structures which characterise rural areas and the difficulties of distance, mobility and access. The centralisation of services, including education, health, retail and advisory services, was offset for some by digitalisation while others (unable to access or to afford broadband) found these essential services even less accessible than before. Some were able to continue their desk-based employment from home, while others were laid off or furloughed. As in urban areas, therefore, inequalities within rural areas were often exacerbated – but sometimes in different ways.

The cuts to council budgets over the last decade have curtailed the provision of public services in rural and urban areas alike. In rural contexts this has intensified the challenges of distance, mobility and access, as services have been withdrawn (many bus routes) or centralised. At the same time, costs have risen as rural populations are older, on average, and populations more dispersed. During the pandemic, this presented further obstacles for those with less mobility and reach. Rural residents without digital access at home were particularly disadvantaged by the closure of public spaces where digital access might normally have been gained, such as libraries, cafés, village halls, GP surgeries and the premises of voluntary and community organisations, in contrast to the widespread availability of Wi-Fi in towns and cities. However, the national lockdowns spurred innovations in the digital economy, e-Health and online education, which may help to mitigate some of the impacts of local service closures on rural communities, now and in the future, as long as digital inequalities are addressed.

A wide range of formal and informal groups across the public and voluntary sectors have provided support to individuals experiencing financial hardship during the pandemic.[3] A positive outcome of the pandemic has therefore been an increased awareness of people living with food poverty and many new community groups have appeared in towns and villages across our study areas. These groups have been lauded for offering a different service from a traditional foodbank, often intervening before people reach a crisis point (Kelleher, 2021). These groups give people different 'entry' points to the welfare system and other support structures, depending on their individual networks. To maintain this support and awareness, it is increasingly important that service providers and the voluntary sector in rural areas continue to play a joined-up signposting role, connecting their clients with information and advice. The importance of voluntary and community action and support was heightened during the pandemic, often filling the gaps left by the inability of the state to reach effectively into rural areas unless in partnership with voluntary and community organisations. As other research in Scotland has found:

> The responses of community organisations and networks have been highly significant as well, with widespread evidence of rapid community engagement and participation, volunteering and generosity expressed in practical action to help the most vulnerable and at risk in particular. [This demonstrates] the power, efficacy, and responsiveness of localism. Resourced, enabled local organisations, which were trusted to respond directly to local needs, were able to act in a targeted and dynamic way. Rural responses to COVID-19 [have been] a collective, whole-community effort. (Scottish Rural Action, cited in Fisher, 2021)

These precariously funded organisations responded quickly and flexibly, despite many of their regular volunteers having to shield because of their age, replenished by new younger volunteers temporarily not occupied at work, and despite loss of their income from charity shops and fundraising. Those voluntary and community organisations with income-bearing assets of their own, such as wind farms or affordable rented housing, had more freedom of action and more resilience in these circumstances. Nevertheless, many face greater uncertainty over sources of funding into the future because of the pandemic. Above all, the pandemic response has signposted how the 'local state' requires the relational skills of the voluntary sector to provide a sustainable response to the needs of vulnerable populations experiencing financial hardship. Developing a new strategic and complementary relationship that 'fully engages locally embedded voluntary organisations at all stages of emergency response and resilience planning' is needed going forwards (Bynner et al, 2021).

As the long-term economic impacts of the pandemic unfold, the findings shared in this chapter reinforce the importance of diversifying rural economies that rely heavily on tourism and hospitality, and of promoting 'good work' which offers a reasonable, secure income. There is also the need for place-sensitive policies, strategies and support for rural communities, a finding that has been echoed in other research during the pandemic (Couch et al, 2020). There are also strong calls for a rural recovery that builds on some of the positive aspects of community responses, including increased levels of local volunteering and opportunities to boost employment and training opportunities (Plunkett Foundation, 2021). Finally, there are signs that the pandemic, and growing familiarity with working from home, are bringing a change in residential preferences with many city dwellers looking to move to rural areas from which they can work permanently or predominantly from home. This 'rural shift', if it materialises, could bring new life, employment opportunities and services to rural areas, although it could also have some potential downsides such as rising house prices and social polarisation. If such a shift is to work to the benefit of rural society, it is important that affordable rural housing and the necessary economic and social infrastructure are provided in good time. These challenges and the associated policy opportunities are explored in more detail in the next chapters.

Changing sources of support: precarity, conditionality and social solidarity

The previous four chapters have examined how changes in each of the four systems of resource allocation (markets, state, voluntary and community organisations, family and friends) affect individual financial hardship, wellbeing and vulnerability in rural Britain. The empirical data presented thus far raises important questions that require further consideration, particularly in relation to the interactions between the four different systems of support. Do they complement each other, or do they compound social exclusion? For example, does support from the state compensate for loss of income from employment? Are there individuals or social groups who might be neglected by several or all sources of support? Do these interactions differ between places? And does the strength of support vary from place to place? We now seek to dig more deeply into the processes of social exclusion which underlie financial hardship and vulnerability in rural places, drawing on our rich data to compare and contrast the three case studies and situate our work within the literature reviewed at the outset.

Precariatisation and the evolution of rural economies

Rural economies have undergone considerable structural change in recent decades, even though this may have been less dramatic and visible than the loss of heavy industry in post-industrial urban areas. The gradual loss of employment in agriculture and forestry exceeds the more intense loss of jobs in the British coal industry, for example. Meanwhile, there has been a growth in service sector activities which now dominate employment in both rural and urban areas. These structural changes were evident across the Western Isles (affecting Harris), Perthshire (affecting East Perthshire) and Northumberland (affecting the North Tyne valley), and in each region there has been an increasing reliance on employment and self-employment in tourism and hospitality, alongside growth in public sector employment (until austerity policies reversed this latter trend from 2010). In conjunction with a decline in crofting, Harris has experienced a considerable increase in 'destination tourism', which has enabled several major new employers to create new job opportunities in the area. However, the growth in tourism in Harris presents a double-edged sword: some employers face staff shortages, partly because of a decrease in the working-age population and partly due to a lack of local affordable housing for employees. In the North Tyne valley tourism has also brought several new benefits, partly counteracting the decrease in agricultural and forestry jobs. In East Perthshire, where tourism is less dominant than in other parts of the county, fewer people are now employed in the previously buoyant soft fruit, textiles and manufacturing industries, and a growing number of people commute to work out of the area.

Despite the upturn in tourism, the unpredictable nature of incomes remains a common feature of rural working life. In all three study areas, many people do not have full-time, permanent employment and there is a tendency for people to have several jobs or work casually. National statistics show that low pay is more prevalent and more persistent in rural Britain than in urban areas (Vera-Toscano et al, 2020). Our interviews confirm that this insecurity is a widespread source of financial vulnerability, with volatile and unpredictable incomes from seasonal/casual work and zero-hour contracts characteristic not only of land-based and tourism employment but extending across many sectors of rural economies. The tendency to engage in precarious, low-paid, seasonal employment was illustrated by the experiences of 'loose farmers' in Northumberland, fishermen in Harris, berry pickers in Perthshire, self-employed creatives 'living on thin air' in Perthshire and workers in hotels and restaurants in all three areas. This precariatisation (Standing, 2011) in rural labour markets has implications in turn when seeking support from the state, as we will see later.

In the North Tyne valley and Perthshire zero-hour contracts and agency working were thought by our participants to be more common than in urban areas, as was the decision to resort to casual work. In Northumberland it was suggested that having more than one job is particularly prevalent among those with no further or higher education qualifications. Although employability and skills issues may limit the potential for people to enter higher-paid jobs, this is only part of the picture. Across the case studies, participants noted the limited range of employers and the impact this has on career progression. For example, employees in rural small to medium enterprises (SMEs) are likely to need a range of general and transferable skills rather than specialist experience. In East Perthshire many job opportunities are deemed unskilled, insecure and unattractive, while in Harris a lack of professional jobs makes it difficult to attract young people (back) to the island. However, there is more optimism in Harris than in the other study areas about future economic potential. Participants talked about renewable energy and marine tourism opportunities and the increased sense of confidence linked to community acquisition of land and assets in the area. Marketing of the 'Harris brand' also presented potential opportunities for sustainable economic growth.

A lack of affordable, flexible childcare compounds the problem when seeking work. For those on low, unpredictable incomes, childcare provision and affordability, often combined with limited public transport to reach distant childcare providers, present a vicious circle in all the study areas. A lack of local and flexible childcare is a barrier to women looking for employment, yet low pay and the costs of travelling to work may make the cost of childcare unaffordable for those who find work. Although there are some positive stories across the case studies of employers offering flexible working hours so that staff can accommodate childcare and other needs, employment only within school or nursery hours remains uncommon. This conundrum appears to make it more likely in rural areas that people with young families have multiple jobs so that parents can work at different times of the day to accommodate childcare and other caring needs. Self-employment may offer a route to more flexible working and although Harris and Blairgowrie are quite buoyant for new business start-ups, this is also not without challenges. These include the cost of business accommodation (where required), digital connectivity for effective working from home and the lack of confidence/knowledge (noted particularly among women in Northumberland) when faced with the prospect of self-employment.

Conditionality and state support

Apart from earnings from employment, people receive support from the state. This comes in many forms, from central and local government, and much has been written about how this has changed in recent decades (for example, Stoker, 1998; Peck and Tickell, 2002). One aspect has been the change from government to *governance*, by which is meant a shift from state sponsorship and provision of economic and social programmes towards the delivery of these through partnerships, with a new role for the state as coordinator, manager or enabler, rather than as provider and director. Such partnerships are now common in rural governance, and examples in our study areas include Community Planning Partnerships (CPPs), health and social care partnerships, local enterprise partnerships and LEADER Local Action Groups. Another change has been neoliberalisation, initially as 'roll-back' (shrinking the state and the institutions of welfarism through privatisation, deregulation and cuts to public expenditure), followed by 'roll-out' of neoliberalised forms of governance and management (re-regulation, competition and tendering, targets, audit, performance management and other new forms of centralised control through subtle and opaque managerial technologies). These changes affect every aspect of state support, including education, health, housing and welfare provision, and extend also to an altered funding context for voluntary and community organisations who must increasingly engage in competitive tenders for insecure contracts and project funding, while adopting the language of business management.

Previous research has shown that claimant rates among those eligible for welfare benefits in rural areas are lower for various reasons, many of which were explored in Chapter 2. Our study found evidence to support a number of the explanations put forward in previous studies, namely that in rural areas fewer of those eligible are social housing tenants who would receive relevant information and support from their landlords; accessing advice and information in distant urban centres is problematic; there are stronger cultures of independence and self-reliance in rural areas, allied to different subjective assessments of poverty and hardship; and there is more visibility and less anonymity in small communities with more potential for stigmatisation of receipt of certain benefits. This latter point was often couched in terms of people's pride, embarrassment, shame or privacy, particularly among older people, and the stigma deterring people from claiming benefits such as Pension Credit, attendance allowance and carer's allowance was a strong theme in all three study areas. This is important because it frustrates the potential for state support to be available at times when income from employment is reduced or lost.

In contrast, there was no stigma attached to receipt of the state pension (although there was some lack of awareness at times), no complexity or conditionality (other than raising of the pension age) and its real value has increased over the last 25 years with a triple lock in place to ensure pensions are inflation-proofed. As a result of this (and the introduction of Pension Credit), poverty among older people fell between 1996–97 and 2004–05 in England's most rural districts from 27 per cent to 21 per cent and in other rural districts from 26 per cent to 18 per cent (DEFRA rural statistics, 2007, cited in Shucksmith, 2008). In our study areas, too, older people were better protected from poverty by the state pension (and other benefits such as Pension Credit and attendance allowance if they claimed them) than in the past. For most older people, then, the state pension is successfully supporting financial wellbeing in old age in the absence of any private pension derived from previous employment. Nevertheless, gatekeepers told us that a significant number of older people in these rural areas are living 'frugally' and often experiencing fuel poverty and a degree of social isolation, with the Citizens Advice Bureau (CAB) in all areas receiving more requests than in the past from older rural residents.

Welfare benefits have provided vital support to rural residents during their typically shorter spells of unemployment or low income, so alleviating financial hardship and mitigating financial vulnerability. They are even more necessary for those facing longer spells of low income due to chronic physical or mental ill health. Recent analysis of the British Household Panel Survey (BHPS) has shown how welfare reforms from 1999 to 2008 were effective in lifting many vulnerable people out of poverty in rural Britain (Vera-Toscano et al, 2020). More recent welfare reforms have sought to reduce public expenditure by limiting eligibility, introducing further conditionality, sanctions and delays and reducing real benefit levels (except for pensions), while gradually transitioning to a new system of Universal Credit (UC). Our research has revealed several ways in which these reforms unnecessarily create financial hardship and vulnerability for many rural residents and reduce the effectiveness of welfare benefits in supporting people in times of need.

Across the case studies, participants told us about the generic challenges faced by welfare claimants. Many of these are not unique to rural residents: the complexity of the online system and the flaws in its design, payment delays, unpleasant experiences at medical assessment centres and so on. However, some issues are compounded for rural claimants. Our participants drew particular attention to the inability of the benefits system (both legacy benefits and UC) to deal fairly with the volatility and irregularity of rural incomes. This incapacity of the system to cope with income volatility was frequently identified as a serious cause of financial hardship and vulnerability in all our case studies, not only because it makes household budgeting hard but also because it increases the risk of debt and destitution. Particularly from conversations with CAB and foodbank staff, we were acutely aware of how volatility in earnings often leads to overpayment of benefits which is then clawed back too rapidly for low-budget households to withstand. Similar problems arise for people in each place from a mismatch between paydays and assessment periods, which can create artificial apparent volatility in incomes, taking people temporarily out of UC eligibility and necessitating reapplication and several weeks' delay when in fact their wages were unchanged. These flaws in the operation of the welfare system therefore exacerbate labour market insecurity and vulnerability, rather than relieve it.

A lot of these challenges are the reasons why people get into debt and highlight how the rural context and cost of living present unique challenges. Most notable across the case studies was the role of delays built into UC pushing people into debt, the impact of waiting for incorrect assessments to be overturned on appeal and the danger of benefit overpayment arising from unpredictable incomes. The close association between debt and poor mental health was also very apparent. In Perthshire debts were referred to frequently in our interviews, either because of shortcomings of the welfare benefits system and/or as following from job loss or marital break-up.

Rural claimants are also more likely to be disadvantaged by the centralisation and digitalisation of the benefits system if they lack access or can't afford access to broadband, broadband is of poor quality, they lack digital skills or because of literacy or mental health challenges. For such groups, the very obstacles or disabilities which prevent them working with the predominantly online and otherwise centralised system may also form barriers to their receiving state support, compounding their financial hardship and vulnerability.

The cost of broadband and/or mobile phone contracts may also be unaffordable for people facing financial hardship and this issue is likely to continue to be in sharp focus as more and more services move online ('digital only' or 'digital by default'). In that scenario, those with limited access to broadband via a computer, or mobile data via another device, or without the funds to afford a contract, remain at a real disadvantage. Our interviewees in Harris felt that those most likely to be affected are the elderly, school children, people on benefits and people without their own transport. This issue is compounded in rural areas by the closure/reduced opening of venues with public access to computers, such as libraries. This is the case in Bellingham and Blairgowrie, where library hours have been significantly reduced in recent years.

Mobile phone and broadband coverage are critical issues for rural populations and interviewees described challenges in all three places. The proportion of homes without superfast broadband is 46 per cent in the Western Isles, 24 per cent in Perth and Kinross and 13 per cent in West Northumberland (Glass and Meador, 2020). Limited coverage of both broadband and mobile phone signal is felt to add to social isolation in Harris and the North Tyne valley, and there is also an issue with access to devices. Although access to IT facilities is better in Harris now than it has been in recent years, connectivity continues to present very serious issues that are 'holding the place back'. For those without broadband, or who lack the digital skills or the necessary income to afford a broadband or mobile data contract, how far those organisations offering benefits advice and support can 'reach' into rural areas remains a challenge, albeit one that professionals are aware of and working hard to address. For example, the CAB's enhanced phone service was praised by participants in all study areas because clients based in remote areas can access advice without needing to travel.

The CAB and other organisations offering advice are also generally working hard to link up to one another, referring people to the most suitable services as required. However, there are still concerns about vulnerable people with literacy problems and/or poor mental health who are no longer able to access face-to-face support via designated outreach visits or drop-in services that have declined in all the study areas and closed completely during the COVID-19 pandemic and lockdowns. These individuals require considerable face-to-face support to complete the forms and to maintain their online journal or they may lose benefits and face sanctions (with knock-on physical and mental health impacts). This support is even more necessary when appealing against adverse decisions, with advocacy and help from voluntary and community sector organisations often particularly important. In all the study areas, for example, the CAB had successfully supported the appeal process for UC claimants and CAB energy advisers had also supported people to challenge incorrect/inappropriate household energy bills and charges.

A final issue mentioned several times in all the case studies is the benefit system's treatment of claimants with a long-term physical or mental illness or disability, several of whom we spoke to. Frustrations were expressed about the generic issue of needlessly frequent reassessments and disregard of evidence supplied by GPs and consultants. A specific rural dimension to these issues is the distance that claimants are required to travel to attend assessment appointments. In all three case studies we heard about the challenges this brings for rural residents. In the Western Isles, there is no assessment facility for those required to undergo a Work Capability Assessment (WCA), with those in Harris experiencing a long delay (often up to a year) before an Atos officer visits the islands, unless they can attend an assessment in Inverness or the Isle of Skye. Even in rural Northumberland, claimants in the North Tyne valley must travel between 30 and 55 miles each way to Newcastle for assessments, often with public transport unavailable from much of the study area. In some instances, these challenges were cited by participants as reasons for delay in benefit receipt or lower uptake of benefits in rural areas.

Apart from central government's welfare provision, the local state also provides important support in many ways to rural residents, including education, health and social care, housing, childcare and many other public services. All of these have been subject to neoliberalisation, especially in England, through cuts to council budgets, privatisation, targets, performance management and conditionality. These have tended to necessitate centralisation or loss of services especially, but not only, following privatisation, despite the best efforts of many rural councils to maintain rural services. Other research has shown a marked decline in private and public services in 'sparse' rural areas of Scotland since 2010 (Wilson and Copus, 2018), and contraction of many rural services is also apparent in England (Rural England CIC, 2022). Our research found evidence of the financial pressures facing councils, health boards and other public service providers and of the consequent centralisation of, or loss of, services; but we also found examples of innovative approaches to rural service provision, often through partnership with voluntary and community sector organisations and joint working, such as the Ageing Well Network in Northumberland and Harris Development Ltd.

Concerns were also raised in all study areas about available support for elderly people requiring social care. Although there is general recognition that the social care system is under great strain across the UK, and notwithstanding better pay and conditions in Scotland, these strains are amplified in all the rural study areas due to greater distances that care workers need to travel, staff shortages and the higher costs of provision. While interviewees in Harris reminisced about times when family, friends and neighbours provided social care informally, there is an increasing reliance on formal care provided by public, private or voluntary organisations, partly as a consequence of the ageing population. Whether in care homes or as care in the community, funding challenges under austerity, combined with the higher costs of delivering formal care in rural areas, have had a negative impact. In Harris and Perthshire staff recruitment and retention present problems in delivering rural social care. In East Perthshire it was suggested that the complexity of the social care system often hampers people knowing that they can choose how their care and support is delivered. Despite a generally negative narrative surrounding social care in our case studies, there might be untapped potential in rural areas to deliver a more personalised and joined-up approach from informal (unauthorised) cooperation between health and care workers.

Local support and social solidarity

The voluntary and community sector has adapted to this changing landscape in various ways. Cuts to public expenditure often reduce the funding available to support this sector while leaving them to try and fill emerging gaps in state social provision. Neoliberalisation increasingly requires voluntary and community organisations to act competitively, perhaps hindering collaboration, to seek income from the state through tenders and contracts, to pursue targets and deliverables perhaps tangential to their purpose and to adopt the modes and practices of private business to pursue others' agendas. An important question is how far such asymmetrical power relationships limit the practical potential for localised and joined-up action. Many of these elements were evident in our study areas, with core funding difficult to find and an increasing need for most voluntary and community organisations to tender for public sector contracts or project funding alongside other fundraising activities. These organisations find themselves speaking a new language, cast as 'customers' working with 'account managers' to prepare 'growth plans' or business plans, for example, or developing new income streams, perhaps in social prescribing where funds still exist. Even in health and social care, voluntary and community organisations spoke of receiving strategic missives from 'Gold Command'. However, this is not universal and many organisations (and council staff) have found ways of resisting or mitigating these tendencies. In all three of our study areas, local authorities were continuing to support CABs and their independent advice services, for example, as well as foodbanks. Moreover, several voluntary and community organisations (notably community trusts) had adopted strategies to draw revenue from community assets such as wind farms or social housing to assure their independence and sustainability, and this gave them greater freedom to act at their own initiative and to respond to local needs.

Our research shows that the support and advice offered by the voluntary and community sector is valued highly by, and invaluable to, those in rural areas experiencing financial hardship or vulnerability. Indeed, we heard that these organisations are most people's first port of call in hard times because of their kind, compassionate and empathetic approach, regarded by some as their *only* potential source of support. Thus, the advice and support provided by CABs was of crucial importance to people needing to claim the state's welfare benefits, or to appeal against adverse decisions or sanctions, given the obstacles and complexity summarised earlier.

Foodbanks have also become an important source of emergency support in all three case study areas, mainly for people of working age who have been let down by the state's system of welfare provision. Across the case studies, many visitors to the foodbank need to access the service because of mental and/or physical health problems or literacy problems which limited their ability to draw support either from employment or from the welfare state (in Northumberland this accounted for over 40 per cent of users in 2017–18). A substantial proportion also had to recourse to foodbanks because of delays in receiving welfare benefits, or as a result of repayments or imposition of sanctions. Since the beginning of the COVID-19 pandemic, the West Northumberland foodbank in Hexham reported that a quarter of the people contacting them were new: 'some have lost their jobs; some are on furlough and struggling to make ends meet and some are self-employed people who are unable to carry on their business in lockdown. Many people who call us live alone and feel very isolated' (West Northumberland, 2021, 2). The challenge of delivering food across their large rural area were immense: deliveries to Kielder involved a round trip of 78 miles.

A range of specialist voluntary and community organisations supported specific groups in relation to poor mental or physical health, housing and homelessness, ageing and social isolation, social care, fuel poverty, debt and domestic violence, among many other challenges. We were impressed by the proactive ways in which many such organisations worked together to complement one another, to signpost people to the most relevant services and to provide integrated and responsive advice, albeit sometimes with time-limited project funding.

It was quite striking just how many third-sector organisations are offering various forms of support, despite the difficulties that these organisations face. This is consistent with the findings from the 2019 Social Enterprise Census in Scotland (Social Enterprise in Scotland, 2019) which reveals much higher presence of these organisations per head of population in rural areas compared with urban (20 per cent of all social enterprises for only 6 per cent of the population). From foodbanks to advice services, from women's refuges to care providers, from fuel poverty initiatives to mental health support, all the representatives that we spoke to noted a range of challenges they face when providing services across large, rural areas. The 'community hub' model adopted by the county-wide network of 'Warm Hubs' and 'Employment Hubs' in Northumberland appears to play an important role both as public spaces and in providing opportunities for specialist voluntary and community organisations to reach beyond their existing clientele and to work together with partners. In Harris community hubs are being considered as a possible model for future coordinated and locally based service delivery across the sectors, and other research has identified the importance of community hubs in these respects, recognising their importance in the COVID-19 rural response (Coutts et al, 2020).

Our research confirmed that rural voluntary and community organisations have become increasingly reliant for volunteers on retired people, who are often highly motivated to help neighbours in their local community rather than good causes at a distance. Despite the strong commitment and motivation of volunteers across these organisations, there are growing pressures of rising demand and limited capacity. This is an issue for the foodbanks in Stornoway, Perth and Hexham, each of whom do their best to serve large areas of dispersed population and irregular public transport. While retired volunteers bring diverse skills and backgrounds as well as high motivation to help others in their communities, some organisations were concerned at the lack of younger volunteers to succeed them as they grow older. This became a much more immediate concern with the onset of the COVID-19 pandemic and lockdowns, when many volunteers had to withdraw and socially isolate because of their greater susceptibility to the virus; but happily, many new, younger volunteers came forward to help during the first lockdown and it is hoped some will continue to volunteer after returning to work.

Although it was not entirely clear to what extent these growing pressures are specifically rural, being able to deliver face-to-face support in rural areas is likely to be dependent on the extent of local volunteer support, the availability of suitable spaces to meet/deliver services, strategies that enable support to be delivered without obvious visibility because of the fear of stigma and effective networking between organisations to identify and support vulnerable individuals and households. Interviews with gatekeepers in all the case studies noted this last point. Examples of successful partnership working include Tighean Innse Gall's (TIG) fuel poverty and home insulation team working with Stornoway CAB to assist benefit uptake in Harris, and through the professional energy advice service provided by the HEAT project in Perthshire. Unfortunately, in March 2022 TIG announced the closure of its home insulation service with the loss of 14 jobs, and a further 16 to 20 jobs in the local supply chain, because of what it sees as the failure of the UK and Scottish governments to island-proof new regulations on retrofitting home insulation. 'The overall effect in adopting these standards when set in the context of COVID-19 restrictions has been to devastate the local supply chain, increase costs massively and halt all installs' (TIG, 2022). For all its success in delivering home insulation and innovation in promoting benefit take-up in an area with the highest rate of fuel poverty in Britain, this experience is a reminder of the vulnerability of the voluntary and community sector in rural areas to decisions taken far away by those without any awareness of their unintended consequences for rural citizens.

Despite their invaluable contributions, voluntary and community organisations face severe funding pressures, making it harder for them to provide services across huge rural areas, and many worried that they will not survive in the medium term. In many cases outreach services have been curtailed or withdrawn for lack of resources, with attempts to provide access online or by phone instead. For many rural residents, access online or by phone is acceptable or even preferable to travelling for face-to-face support but, as noted previously, for many others it is unhelpful or unfeasible, notably for people who lack literacy or digital skills and require face-to-face assistance. A plurality of means of access to help and support is important, not only because some people do not have digital access or skills, but also because some vulnerable people may require face-to-face support for reasons of literacy or mental health.

Family, friends and neighbours are another important source of support in all of the study areas, but people's ability to draw on this source varies considerably according to both the characteristics of particular place communities and social norms, social capital and personal relationships. This type of support is particularly strong in Harris, although some participants noted that this has dwindled somewhat more recently in scenarios where both members of a household are working, or in the more common choice of young people to leave the island, and/or the in-migration of new people to the islands for whom it takes time to build up networks. However, the strength of this type of support re-emerged during the COVID-19 pandemic and lockdowns. In East Perthshire there is a strong reliance on family members for care at home because of the challenges associated with recruiting care workers in rural areas. We heard less about this type of support in the North Tyne valley. Instead, participants told us repeatedly about strong community spirit and a culture of acceptance in parts of the area, and that people are generally aware of what their neighbours need. There was a sense that 'people would notice if there was an issue', and that anyone would routinely pick up shopping for other residents.

There is a widespread idealisation of rural communities as places where everyone looks after one another, and there was much evidence from our study of the importance of neighbours, friends and families in smaller communities. This may be associated sometimes with the stigma associated with claiming benefits leaving rural dwellers more heavily reliant on friends, family and social networks for support. Either way, as these social networks decline, these individuals and families are at even greater risk of falling into deeper, more persistent and even more hidden poverty (see Lichter and Graefe, 2011; Tickamyer and Henderson, 2011; Sherman, 2013). Moreover, the strength of community networks and support varies considerably from one place to another. This unevenness of informal support also occurs at the scale of individual place communities within each study area, with some villages and small towns more tight-knit and characterised by informal help than others. Even in normal times, such social networks were important in learning about opportunities for employment or for rented housing, or with informal childcare, and (if they could afford it) parents routinely helped sons and daughters find the deposit to buy a car or house. In hard times this relationship might be more complex: we heard that people might not request help, nor accept help if it was offered clumsily, but that help could be offered and accepted if this was framed in accordance with social norms. For example, people might not admit to being hard up, but would accept lamb chops from a neighbour who had killed a lamb. While this demonstrates people's resilience, it also reveals how this can make disadvantage hidden and how it might be hard to access local support networks for those who do not understand local social norms and lexicons or who have not made the social contacts. On top of this there remains a very real stigma attached to seeking 'charity' or to those who are perceived to have transgressed social norms in one way or another.

Individual coping strategies and agency

Within local and regional opportunity structures, individuals and households displayed a repertoire of strategies to draw support from the various systems and to avoid financial hardship. At one level, the most common strategy was to budget carefully, buy from jumble sales or charity shops, if necessary, save when one could and avoid debt – a more difficult task when incomes are low and irregular, volatile or seasonal, but also challenging when those on middle incomes face redundancy and have to adapt to a lower income.

At another level, we found people following conscious or unconscious strategies to find employment, or additional/better employment; to access pensions and other welfare benefits; to receive help, advice, companionship and perhaps food through voluntary and community organisations; and to gain help or retain social capital among their immediate circles of family, friends and neighbours. Sometimes these conflicted with one another, for example, if accessing a foodbank or welfare benefits brought social stigma. Often, though, they worked in tandem as when casual employment was found through social networks, or when advice and support from voluntary and community organisations assisted in applications for welfare benefits.

Effective strategies had to be appropriate to the context. Thus, while well-paid, secure employment was typically found through formal job applications following public advertisement of the vacancy, local work was more likely to be found through word of mouth or social media and might not be widely advertised. An effective strategy for finding work, as well as other support, would therefore include building and maintaining strong social networks, both face to face and through Facebook and WhatsApp groups, while participating in the 'right way' (respecting social norms) in community activities, perhaps through a faith group, local school or village hall. Another effective strategy might be to cultivate a relationship with a gatekeeper in the council, social landlord or advice service so as to learn more about the benefits system and receive help in maximising such income: this might risk social stigma, if perceived as a 'scrounger', but would be less risky for a social housing tenant or an individual recognised to have physical or mental health issues.

It was clear in all case studies that budgeting can be challenging, with emergency fuel top-ups, for example, often linked to the wait for the first UC payment to come through. In East Perthshire there were very few mentions of savings, except in the context of private and state pensions. It was noted that, in contrast to those who retire to the area from professional jobs with employment-related pensions, few people in rural occupations are likely to have similar pensions and the main assets which people identified are secure homes and cars. Sole operators and microbusinesses were also thought by gatekeepers to make no pension provision and therefore to have to work beyond retirement. Those with council house tenancies in East Perthshire regard themselves as fortunate, especially where their rent can be paid directly from Housing Benefit so that they have no worries about falling into rent arrears. Such tenancies are important assets.

When asked how they coped with unexpected bills or debts, respondents described various strategies and sources of help. In general, although payday loans had been turned to in some cases, these were universally seen as carrying considerable risks and storing up problems. In Harris older people were thought to be less likely than younger people to take out loans and be in debt. The credit union in Stornoway was well used by older people on a pension who wanted the ability to save a small amount on a regular basis. In Northumberland a community bank offers an affordable loan scheme, which is used mostly by people in work, highlighting the extent of in-work poverty as a result of low pay and volatile incomes.

There are numerous sources of help and support available to residents of all three case study areas, although many are located at a distance in Stornoway (for Harris), Perth (for Blairgowrie) and Hexham (for the North Tyne valley). However, despite this range of provision of support, rural people often had difficulty in finding appropriate help. Again, this highlights the very real challenge of 'spreading the word' effectively in rural areas. In Blairgowrie, even when individuals found a source of support, they found it frustrating and difficult if they spoke to a different person each time, sometimes without knowledge of the area and of the rural context, and perhaps passing them on to other departments. It was much preferred if a personal connection could be established and trust built up.

People's individual strategies and their capacity to improve their situations also play a role. In Northumberland one interviewee noted that people experiencing financial hardship in rural areas may be less able to "take risks" to improve their situation than their urban counterparts. In Bellingham there was a strong reliance on a regular jumble sale for sourcing (pre-owned) luxury goods locally. In Perthshire one individual would not claim welfare benefits on principle, devoting their energies instead to finding sources of casual work and maintaining the personal networks necessary to this, while reducing their expenditure by turning off central heating and self-provisioning. In practical terms, a common and effective strategy suggested by gatekeepers was learning how to budget more effectively. However, this was not seen as straightforward and there appears to be potential for more support in this regard.

Intersectionality: are sources of support cumulative or complementary?

Looking at all four sources of support and how they act cumulatively to offset or reinforce social exclusion and financial vulnerability, we can begin to draw some conclusions. People with high educational qualifications and access to secure, better-paid employment in urban areas are favoured by markets, whereas those lacking such credentials working in many rural occupations and local workplaces, and people experiencing physical and/or mental illness, are relatively disadvantaged by markets. This is not only a matter of rates of pay but also of working conditions, job security, pension potential and perhaps job satisfaction, with worsening job insecurity, uncertain hours and in-work poverty experienced in rural workplaces. In turn, this precariatisation of employment and the associated 'low escape rate' from low pay (Shucksmith, 2018a) creates precarity in relation to housing, food and many other essentials of life.

In principle, the welfare state should offset this precarity, providing a safety net in hard times and ensuring 'social security'. The welfare state's support has assisted those with insufficient income from employment, but while pensions have grown in real terms in recent years, overall welfare spending was reduced in real terms by £26 billion between 2010–11 and 2016–17, with further planned cuts taking this to £40 billion by 2020–21, prior to the COVID-19 pandemic (Keen, 2016). A series of welfare reforms has reduced the real value of benefits, while intensifying work activation and conditionality: the cumulative impact has been a redistribution of social and societal risk towards the most vulnerable, with young people, people with mental or physical illness and lone parents particularly disadvantaged.

Thus, while state support still offsets disadvantage for those who can navigate the complexities of the system successfully, for many others welfare conditionality and cutbacks mean that state systems now often compound the precarity of increasingly flexible and insecure labour markets. As we have seen, risk is individualised and social solidarity diminished, in rural areas as in urban. During the COVID-19 pandemic, the state did step in to provide vital, temporary support to many employees and to businesses, although around 3 million people were ineligible for one reason or another (Kempson and Evans, 2020), and raised UC temporarily by an additional £20 a week (Brewer and Handscomb, 2020) until this was withdrawn again in September 2021.

Meanwhile, the challenges of maintaining and delivering services in rural areas had already been heightened by cuts to English local authority budgets over the decade prior to the pandemic. The National Audit Office (NAO) (2021) found this fall of around a third in councils' spending power, alongside rising demand for services, had left councils more vulnerable to the impacts of the pandemic. The NAO also warned of continuing cuts to services in the next few years, including social care, special educational needs, libraries, buses and community centres, as councils struggle to meet the extra costs incurred during the pandemic: 94 per cent of councils expected to have to cut spending in 2022–23 to meet legal duties to balance their budgets, and several risk insolvency (NAO, 2021). This is likely to lead to further centralisation or loss of services in rural areas. Councils in Scotland have faced smaller cuts under the Scottish government, but they still face budgetary pressures with budgets unable to keep pace with increasing costs, particularly due to loss of income during the COVID-19 pandemic (Audit Scotland, 2021).

Specific rural dimensions to this arise from the volatility of rural incomes, digitalisation and digital exclusion, difficulties in accessing advice and support and typically lower claimant rates. Higher costs of living and widespread fuel poverty are further important aspects. Voluntary and community organisations have been active in supporting many of the people disadvantaged by markets and the state, and their services are highly valued, but again there are challenges of reach into rural areas as well as resource constraints which tend to necessitate further centralisation and digitalisation. Noteworthy, among a wide range of voluntary and community organisations in rural areas, are CABs, as an indispensable source of help in applying for welfare state benefits, and the foodbanks: each of these face growing needs with limited resources and reliance on volunteers, whose availability varies from one community to another. Many voluntary and community organisations seek to transform the lives of those who seek their help, while also responding to urgent need at times of crisis and fulfilling roles which in urban areas are played (and funded) by the state. We consider in the final chapter of this book the extent to which the valuable work of voluntary and community organisations in rural areas might indeed mask and facilitate the failure of the state to provide for its citizens in all places. Nevertheless, some people find it hard to access the support of these organisations, including people distant from urban centres who require face-to-face help or lack digital access, those who live in areas with fewer active voluntary and community organisations and some less visible minority groups.

Family, friends and neighbours are the remaining source of support, most available where there are strong community networks and values. There is a widespread perception that all rural communities enjoy such informal support networks and community solidarity, and again this may encourage service providers to think that less formal provision is needed. Informal support may be harder to access where family and social support networks are stretched or fractured, perhaps by migration or a lack of affordable housing, or where social norms and values are exclusive rather than inclusive.

The interactions between the different sources of support may be illustrated through a few selected experiences of individuals we spoke to in our research. The first story shows how the various systems of support can each play a crucial role at different points in a young couple's life, with family filling gaps when other sources of support temporarily fail, and affordable housing from a housing association compensating for market failure. A self-employed man had taken out a loan to buy essential equipment for his work, but health issues prevented him from working and he couldn't meet the repayments. His mother helped him cover these costs until he was well enough to go back to work, which thankfully he has been able to do, and he has now managed to clear the debts. He and his fiancée rented a one-bed property from a housing association, but they wanted to start a family and needed a bigger home. They tried to get a mortgage for a shared-equity house but his self-employment and her seasonal work prevented this. Luckily, she now has full-time employment, and this has enabled them to make the move.

The second account is of a man in his 50s who lived "up the Glen". He had casual work on the local estates and lived in a rented cottage, but the work had stopped in the winter so he had no income and couldn't pay the rent. This meant he had to travel many miles to the council's housing office on his bicycle as he did not have a car. While there, he would visit the foodbank and carry some supplies back home. During part of the year there would be plenty of work, but when he visited the foodbank there wasn't any: it was a precarious existence. Subsequently, he was evicted from the cottage because the estate wanted to let it as a holiday cottage. This illustrates how insecurity and precarity in both labour and housing markets compelled someone to travel many miles without public transport to seek help.

The third and fourth examples illustrate the complex issues facing people with mental ill health and the difficulties of gaining the help and support they need. A 63-year-old lady with no money moved into council accommodation after family breakdown. She suffers from medical problems and long-term mental illness but isn't classed as high need so has no dedicated council support worker. Although the council's welfare rights team were helpful in providing advice, she needed to phone a number and navigate a menu which is beyond her capacity. So, she needs lots of support even to get to the point where she can access advice, never mind put together an application for benefits. We spoke to her just before the COVID-19 lockdown and we expect she would have repeatedly experienced this challenge due to the complete removal of any face-to-face support. This lady must rely completely on local voluntary and community organisations to help her apply for state welfare and support. If this help were not available, she would be without income or support.

In the final case, a former nurse suffered mental illness and marital breakdown, which led to loss of employment, homelessness and debt. Initially, it was hard for her to know where to turn, but through support from voluntary and community organisations, the council and a psychiatrist, she found a women's refuge, was rehoused in council housing, accessed welfare benefits, resolved her debts and started to rebuild her confidence. Her parents helped with the purchase of a car, primarily so she could access support from voluntary and community organisations and other services. This lady worries about WCAs and needs considerable help with benefit forms and ongoing therapy. She considers herself lucky to have found good support which is helping her on a long road to recovery, but she still needs help from many sources.

In each of these stories it is notable once again that voluntary and community organisations have played a vital role in helping, despite the challenge of insecure and inadequate funding which hampers their capacity to reach beyond the main settlements into rural hinterlands. Where various sources of support complement one another, appropriate support appears to reach people in need more quickly and smoothly than might otherwise have been the case. This emphasises the importance of good communication and signposting between support organisations, combined with support from family and friends to fill any gaps. Getting individuals 'on the right track' to support is an important first step, particularly if an individual lacks agency or capacity (or is too proud/ashamed) to seek help.

The role of place and rurality

The research confirms the role of place as an important dimension of intersectionality, highlighting how place modifies and intensifies the effects of other social characteristics such as class, gender, age and ethnicity. Living in a small community is a feature of rural life which has positive and negative sides, again with different impacts from one place to another and according to social position. As we have seen earlier, positive aspects might include kindness and care from family, friends and neighbours, sharing of information about work and other opportunities, social interaction and practical help in times of need. On the downside, small community life can be subject to social norms, social control, surveillance and potential stigma, limiting people's freedom to live their lives, and operating to include or exclude people and social groups from these (and other) sources of support.

This may be related to social change, loss of traditional practices and values (traditions of help) and challenges to established place identities and social norms. Savage et al (2005, 96) introduced the concept of *local habitus* to refer to 'dominant local lifestyles and shared dispositions, sustained by local neighbouring relationships as well as by common experiences. People's sense of belonging to the area is related to the way that they share these dominant values'. Recent research in Northumberland (Black et al, 2019) found a moral economy of hard work and independence which was often seen as intrinsic to rurality itself and which stigmatised people who claimed welfare benefits or visited a foodbank. Black et al (2019, 273) argued that this discursive framing is an instance of Bourdieusian symbolic violence

in so far as those affected are complicit in this exercise of power over them, accepting a narrative of self-reliance and stigmatising of welfare as self-evidently right. Moreover, the impact of an economic crisis on those already disadvantaged may be greater where such a moral economy stigmatises and restricts access to non-family means of support.

This book shows that this place habitus and symbolic violence exists in other rural areas beyond Glendale, to greater or lesser extents.

An important aspect of this is the social construction of each place, and indeed of rurality itself. Much has been written about how the framing of rural places as idyllic conceals rural poverty, making it less likely that this is recognised or addressed. Similarly, the framing of rural places as natural and requiring protection from development can obstruct the provision of housing or workplaces. Beyond these general discursive framings of rurality, each place has its own specific narratives of change and identity (local habitus in Bourdieusian terms) which inform, enhance and constrain their future potential and have differential social impacts. Moreover, different social groups may have differing constructions of their place which compete for dominance in public and political arenas. In Harris in the 1990s such competing discourses of place surfaced in disputes over the proposed Lingerbay superquarry, for example (Mackenzie, 1998). In later work Mackenzie (2013) shows how neoliberal norms of enclosure and privatisation were disrupted by community land ownership, leading to new, more hopeful narratives, framing the islands instead as 'places of possibility'. A perennial challenge is how to promote such positive narratives while also highlighting need and disadvantage.

In Perthshire, which is generally seen as an affluent county, gatekeepers felt that it is difficult and 'unacceptable' for individuals to talk about financial hardship when they are so visible in their community. In Harris, stigma, visibility, privacy and pride were mentioned repeatedly, particularly in relation to the older generation. The foodbank in Stornoway has tried to reduce the visibility of people using their service, for example, by keeping a supply of parcels in the cars of other support workers doing home visits, to be given out as required. However, questions were raised about whether the inconspicuous location of the Perth foodbank (down a narrow alleyway in the city centre) reinforces feelings of secrecy and shame for those who need help.

Connection between place and 'loss' can also affect positive narratives. Place-based narratives of what has been lost from rural communities were present in all our case studies. In the North Tyne valley and East Perthshire loss of industry and associated activities was a common theme. In all three areas we discussed the impacts of loss of people (particularly young people) on rural communities, as well as the loss of services. Equally, though, narratives of optimism and hope were often founded on a sense of place and cultural identity, notably in Harris but also in other places. Savage (2010) argues that such narratives of loss may partly be a response to the in-migration of more affluent incomers attracted to 'enchanted places' where they can 'put down roots', often identifying primarily with landscape and nature rather than social belonging and deploying very different symbolic and cultural capital. Savage (2010, 33) shows that 'to counter the moral claims of the newly arrived' to these enchanted places, established residents with fewer resources deploy narratives of disenchantment and loss as a form of resistance to gain alternative and distinctive forms of cultural capital through their relationship to place and claims of 'localness' and 'local authenticity'.

The higher cost of living in rural, remote and island areas was an omnipresent theme during our interviews. This was not surprising as it is already well documented that budgets required by rural households to achieve a minimum acceptable standard of living are considerably higher than elsewhere in the UK (Smith et al, 2010; Hirsch et al, 2013; Davis et al, 2016; 2021; Scottish Government, 2021a; 2021b), the more so with the large increases in energy costs since 2021, which have also exacerbated fuel poverty (Scottish Government, 2021b). Those in Harris noted that living costs are even higher in the islands than in rural mainland areas, particularly in relation to island delivery costs, although this is partly resolved by there being fewer activities on which to spend the remainder of a household's income (for example, cinema, pubs and so on). Most of the leisure activities that people take part in on the island are centred around those organised by committees, community events, visiting and so on. The high cost of public transport was also a concern in all three areas, with specific barriers presented by unhelpful timetabling/routes to people without private transport. Fuel poverty was an issue in each of our study areas, affecting a larger proportion of homes than in urban areas.

Access to good work and secure, affordable housing depends upon local employment opportunities and social housing provision as well as on education, skills, mobility, health and support from family, all of which reflect age, gender and social class. Access to public services, whether face to face or digital, is also likely to present challenges: digitalisation may help some but we have also seen how this excludes those unable to afford or to receive broadband or mobile signal, as well as those with mental health issues or lacking digital or literacy skills. Difficulties of distance, mobility and access are particularly severe for disadvantaged groups in remote and island areas. Costs of provision of social care at home are likely to be higher, if staff are available at all. And people eligible for welfare benefits for one reason or another faced barriers of distant sources of advice and help, digitalisation and centralisation of welfare support, inaccessible WCA centres, along perhaps with social stigma.

Our research also shows how different places have different institutional capital, in terms of their knowledge, networks and capacity to act collectively in pursuit of shared objectives and shared values (see also Shucksmith, 2010; Fischer and McKee, 2017; Children's Neighbourhoods Scotland, 2018). The best of intentions needs to be translated into practice to have any effect. Different places also have different institutional capital with which to support one another and to act collectively in pursuit of shared goals and values. Community assets, from village halls and public spaces to revenue from wind farms, enabled people to work together and address needs, especially during the COVID-19 pandemic and lockdowns. Softer community assets, such as leadership, kindness, skills in conflict resolution and connections to public agencies, were also very important in building inclusive communities and enhancing local opportunity structures for the least powerful in rural societies. In each of the study areas there was evidence of the beneficial legacies of past efforts at community development and their long-term building of institutional capacity. Again, this affects the local opportunity structures within which individuals, households and social groups live their lives.

Our three study areas offered very different local contexts and local opportunity structures, while also having much in common. Parts of Eastern Perthshire offered easy access to well-paid jobs in Perth or Dundee, enabling some people to earn high incomes while also accessing urban services. In the remoter study areas, especially in an island context, such opportunities were less available. In Harris the rapid expansion of destination tourism had brought work where it had been lacking but at the expense of housing affordability. Community Development Trusts or community land trusts were bringing innovative approaches to creating new opportunities in parts of Northumberland and in Harris but each of these was working with its own assets and challenges in place-based actions. Governance contexts also differed significantly, with Harris benefiting from more local control since the new islands council, Comhairle Nan Eilean Siar, was formed in 1975 to replace control from Inverness-shire and Ross-shire in Inverness and Dingwall respectively; in contrast, Tynedale District Council, based in Hexham, was merged into a unitary authority, Northumberland Council based in Morpeth in 2018, taking administration and services further away from North Tyne valley residents.

The point is that rural areas offer very diverse local opportunity structures, power relations and policy contexts, and these differences intersect with age, gender and social class to necessitate place-based responses alongside standardised person-based policies. This necessary interaction between place-based and person-based policy approaches is the subject of the final chapter.

Conclusions and policy implications

This study has explored the processes of social exclusion in rural Britain by focusing on changes in the four sources of support available to households in diverse rural localities (markets, state, voluntary and community sector, and family and friends) and how they interact to reinforce cumulatively or to substitute and offset social exclusion and financial vulnerability. This conceptual approach has proved effective in revealing the connections between individual experiences and broader processes of individualisation, precariatisation, labour market flexibilisation, welfare conditionality, digitalisation, 'roll-back' and 'roll-out' neoliberalisation and austerity and the extent to which these are modified by place. The framework has also revealed emergent agency on the part of rural residents (individual and collective) and highlighted issues around civil society and community empowerment.

In conclusion, this chapter offers some closing reflections on the original contributions of this study to our understanding of poverty and social exclusion in rural Britain and on the implications for policy and practice. We begin by reflecting on the main themes emerging from this study and highlight some of the new insights which add to our understanding of social exclusion in rural Britain. We then summarise previous studies' suggestions for policy interventions and go on to consider the potential to combine person-based and place-based policy approaches to social exclusion in rural areas. Building on these, we highlight some of the most pressing, immediate policy challenges and suggest eight practical opportunities for policy development to address these and promote social inclusion in rural Britain. Finally, we share some closing reflections relating to issues of power and governance.

Main themes emerging and original contributions

While statistics show that average incomes are somewhat above the national average in accessible rural areas, this research confirms the findings of many previous studies that poverty, financial vulnerability and social exclusion affect many people in rural as well as urban Britain, but that in rural Britain it is less visible and less likely to be addressed by policy. As noted in the introduction, half the households in rural Britain fell below the poverty line at some point during the 18 years between 1991 and 2008, at which point the impacts of the financial crisis and subsequent austerity will have increased vulnerability further alongside cuts to public services and social welfare.

One of the most striking and original contributions of this new research is the detailed analysis of the way the welfare system operates in rural Britain and the challenges this brings for the poorest and most vulnerable in rural society. Welfare benefits have provided vital support to rural residents during their typically shorter spells of unemployment or low income, and recent research (Vera-Toscano et al, 2020) has shown the effectiveness of welfare reforms during the 1999 to 2008 period in reducing poverty in rural Britain. Welfare reforms since 2010 have sought to reduce public spending by limiting eligibility, introducing further conditionality, sanctions and delays and reducing real benefit levels (except for pensions), while gradually transitioning to a system of Universal Credit (UC).

Our research has revealed several ways in which these reforms unnecessarily create financial hardship and vulnerability for many rural residents and reduce the effectiveness of welfare benefits in supporting people in times of need. Participants told us about generic challenges faced by welfare claimants, in urban and rural areas alike, such as the complexity of the online system and the flaws in its design, payment delays, unpleasant experiences at Work Capability Assessments (WCAs) and so on. Some issues are compounded for rural claimants, notably the inability of the benefits system to deal fairly with the volatility and irregularity of rural incomes: this was a major cause of financial hardship and vulnerability in all our case studies, not only because it makes household budgeting hard but also because clawback of the resulting overpayments increases the risk of debt and destitution. Other shortcomings included the role of delays built into UC again pushing people into debt, the impact of waiting for incorrect assessments to be overturned on appeal, the inaccessibility of WCA centres and the difficulty in accessing help and support whether face to face or online.

Claimant rates are known to be lower in rural areas than urban, and to fall systematically as settlement size decreases, and this is usually ascribed both to social stigma in claiming benefit entitlements and to distance from sources of advice and information. We found that rural claimants are also more likely to be disadvantaged by the continuing centralisation and digitalisation of the benefits system if they lack access or can't afford access to broadband, broadband is of poor quality, they lack digital skills or because of literacy or mental health challenges. For such groups, the very obstacles or disabilities which prevent them working with the predominantly online and otherwise centralised system may also form barriers to their receiving state support, compounding their financial hardship and vulnerability.

Many rural communities have also experienced marked social and demographic changes as rural homes are purchased by commuters and/or retirement migrants while young people leave to find further education, secure employment and affordable housing elsewhere. These changes bring an older, wealthier middle class to live in rural areas, widening income and wealth inequalities, inflating house prices and disrupting social networks and norms, while also bringing new ideas, knowledge, volunteers, experience and connections of potential community benefit. The contrasts between locals and incomers, rich and poor, were apparent in each of our study areas, whether accessible or remote, accompanied by powerful narratives of social change – often of loss. These changes, and their impacts in terms of widening inequalities, housing (un)affordability and social cohesion, have important implications for people's financial vulnerability and other aspects of their wellbeing, for example, in fracturing support networks, concealing hardship or (more positively) enhancing voluntary action and social innovation.

The fracturing or stretching of informal support networks, along with the social stigma attached to receiving charity or claiming welfare entitlements, can make it very difficult to seek help in hard times. Indeed, this might be seen as a 'dark side' of individualisation in the risk society and of the culture of self-reliance cultivated in rural societies. In this context, the ability of voluntary and community organisations, notably Citizens Advice Bureaus (CABs), and the state to reach into rural areas in sensitive and appropriate ways is invaluable. This challenge is discussed further later.

Community assets also have relevance here, in enabling such action. We found evidence that community assets can be helpful in building community confidence and institutional capacity to act as well as in strengthening and repairing social support networks. In all areas (and particularly Harris), participants discussed the greater role that might be played by community trusts in the future. The independent revenue stream from the ownership of such assets gives community organisations greater freedom to act on their own initiative and to respond to local needs, perhaps extending to joint working with other partners, including local councils, to develop innovative and creative delivery of rural services. But, while taking on additional responsibility and/or developing initiatives to support people experiencing financial hardship may seem desirable, participants in our study were concerned about the extent to which community trusts may take on additional risk and depart from their original strategic focus. Nevertheless, there is clearly potential for creative solutions and support to be delivered by local community organisations in stronger financial positions, as evidenced also by the experience of Glendale Gateway Trust in Northumberland (Healey 2018; 2022).

It may also be helpful to reflect briefly on the methodology employed in this study, and on the experience of undertaking fieldwork during the COVID pandemic. As noted earlier, we were pleasantly surprised by the experience of conducting such sensitive interviews online or by phone, with trust rapidly established and very candid responses to questions. This might not have been the case for some of the earlier interviews with highly vulnerable people, which we were fortunate to conduct face to face before the pandemic. We were also careful to adapt our methods for the online focus groups in line with others' experiences, and these were also very successful. We missed the possibility of meeting one another face to face in research team meetings during the analysis, but again were able to do this online.

One of the challenges of conducting research on a potentially sensitive topic in small communities is our duty of care towards those who generously agreed to speak to us to protect their anonymity and confidentiality. Necessarily, this may make the voices in this book appear somewhat disembodied because we cannot reveal more about the respondents without compromising their anonymity: even the use of pseudonyms could reveal someone's identity by enabling details from different quotes to be amalgamated. For this reason, also, we cannot thank our respondents publicly, as we would wish.

What policy interventions have previous studies proposed?

Because the factors contributing to social exclusion in rural locations often remain hidden, and we lack evidence on their nature and extent, policy interventions are not always well tailored to address them (Milbourne, 2016a). As summarised by Milbourne and Coulson (2020), rural poverty remains hidden from view in a variety of ways: physically, through the dispersed nature of rural settlements; statistically, as poor households tend not to be as spatially concentrated as they are in cities; politically, as policy interventions largely follow the most visible manifestations of poverty; and culturally, as established national discourses of rurality (reinforced by wealthy in-migrants seeking the rural idyll) tend to deny the presence of poverty in rural settlements and local cultural norms in many rural places downplay the significance of both poverty and state forms of welfare assistance.

As a result, as Milbourne (2006) argues, area-based policies or community development initiatives that fail to attend to these specificities of rural poverty are unlikely to be successful. Instead, tackling rural poverty requires the development of interventions that are based on, and respond to, people's lived experiences of poverty in rural places (see also Milbourne, 2014). As Milbourne and Coulson (2020, 4) argue:

> Successful interventions also need to engender dialogue with and the empowerment of local community, rather than be based on unidirectional and top-down processes of community consultation and engagement. What this means is that community-led initiatives to tackle rural poverty need to begin from a position that recognises the importance of everyday lived experiences of low income, crafted in and through time and place. As such, it is critical that the contextually

embedded experiences of poor people inform strategies of community-led initiatives.

Of course, as Milbourne and Coulson argue, such community-led interventions will not be possible in all communities as they will depend on the levels of capacity, assets and skills (including leadership, ability to discuss and prioritise and so on) in those communities. Powell et al (2018) propose four principles that they consider underpin community-based interventions for tackling poverty in rural areas:

- building on the existing capacities, assets and knowledges of rural communities as well as identifying current challenges;
- developing community alliances based on the lived experiences of rural places;
- supporting rural communities with resources, knowledge exchange opportunities, processes of empowerment and nurturing strong leadership;
- developing multi-stakeholder, multi-sector collaborations led by communities to develop strategic interventions that are sensitive to the specific contexts of rural places.

Based on these principles, the authors recommend a three-pronged approach to tackling rural poverty that builds economic, social and environmental and health capital, supports needs and encourages strategic planning and action to promote more integrated approaches. Milbourne and Coulson (2020) present a critique of Powell et al's (2018) work, identifying several issues worthy of further consideration.

It is worth also briefly mentioning other work which has focused on identifying appropriate policy and practice responses to rural poverty. For example, the Joseph Rowntree Foundation's work in 2016 (Prosperity without Poverty) suggests that, while government must take the lead in tackling poverty, businesses, communities, faith groups and individuals, as well as public sector organisations such as local authorities, also must play a part. Cited in Milbourne and Coulson (2020), Crisp et al (2016) argue that attention needs to be given to the material and non-material impacts of community-based intervention; the former might include cost reductions in areas such as housing, fuel and food, while the latter might include reduced social isolation, increased social participation and feelings of empowerment among individuals on low income. The success of such interventions should be judged on outcomes and the process of participation. As Milbourne and Coulson (2020) go on to argue, community-led approaches to tackling rural poverty need to be accompanied by broader, multi-scalar, integrated and strategic policy interventions to tackle structural issues such as economic development, housing, transport and digital connectivity. In relation to service provision, for example, Bailey et al (2004), in their study of deprivation in Argyll and Bute, argued that given the spatial distribution of deprivation that they observed, services or actions to meet the needs of those experiencing deprivation need to be accessible across the local authority area. Deprivation is not so highly concentrated that efforts can focus on only one area or set of areas. At the same time, there are some areas where needs are significantly higher than average, and services in these locations are likely to be under more severe pressure. Local authority resources need to be distributed accordingly and flexibility of response, including through partnership working, is critical (see also Pacione, 2004; Children in Wales, 2008).

More broadly, a change of attitude is required among urban dwellers and largely urban-based policy makers to recognise that poverty exists in rural places and that it manifests differently in these places, in terms of differences between rural and urban poverty, and differences in poverty experiences between rural communities. Allied with this, work to promote awareness and uptake of welfare entitlements among rural people across the different demographic groups is important. Recognition of the higher cost of providing services in rural areas is also required (for example, the longer travel times for social care workers visiting households), with resources provided accordingly (for work on social exclusion in the UK, see also Phimister et al, 2000; Shucksmith 2000; 2004; Dax and Machold, 2002; Jentsch and Shucksmith, 2004; Milbourne, 2004, and for related studies in the US, see Brown and Hirschl, 1995; Rank et al, 2014).

The potential to combine place-based and person-based policy approaches

It is striking how often people's awareness of, and application for, national person-based measures, such as welfare entitlements, has been facilitated by local place-based measures, such as advice and support from CABs and other voluntary and community sector organisations, or through local partnership working. It is also evident that national policies are designed and implemented without the benefit of local place-based knowledge.

People's experiences during the COVID-19 pandemic and lockdowns have provided even more evidence that both national, person-centred measures (for example, the UK Coronavirus Job Retention Scheme (CJRS) and UC uplift) and local, place-based measures (for example, advice services, foodbanks, local institutions, community hubs, project working) can make a significant difference to people's financial vulnerability and their wellbeing. This raises important questions about how policy can enable synergies between these two types of interventions so that they work well together. For example, how might policies not only promote local uptake of person-centred measures, but also allow for local knowledge to feed upwards into better formulation of person-centred measures?

To this end, national policy development could be better informed by local knowledge and local practice from rural areas, not only through the involvement of local stakeholders in rural proofing as policies are being developed and piloted, but also through mechanisms within national government for continuous learning and policy refinement from local experience. As part of such an approach, piloting of new national initiatives might routinely take place in rural localities, with evidence of learning from these attached to policy announcements in the form of rural impact statements. At the local level, organisations such as Community Planning Partnerships (CPPs) could be given a formal role, not only in promoting joined-up working but also in (publicly and transparently) feeding back to national government where further adaptation to rural contexts is necessary and where good practice has been identified and might have broader application. The diagram in Figure 8.1 shows a range of opportunities to support rural people experiencing financial

Figure 8.1: Policy framework of opportunities to support people experiencing financial hardship and vulnerability in rural areas

hardship, spanning the four systems of support and combining person-based and place-based measures. Some of these opportunities are pursued further in the next section of this chapter.

Immediate policy challenges and opportunities

There are some more immediate and practical conclusions to be drawn in respect of policy and practice, and this book challenges current policy and practice in several respects. One aspect of this is the question posed previously of how better to combine person-based and place-based approaches.

The research has identified nine overarching challenges that span the four systems of support (markets, state, voluntary and community sector, and family and friends) and were apparent in the empirical evidence collected in the three case studies. These challenges speak to several UK government departments (for example, Cabinet Office, DEFRA, Department of Work and Pensions; Department for Levelling Up, Housing and Communities, Department for Energy and Industrial Strategy) and to the Scottish government across many of its directorates (for example, Agriculture and Rural Economy, Economic Development, Fair Work, Employability and Skills, Housing and Social Justice, Local Government and Communities and Social Security), as well as many other national agencies and stakeholders, local government, local enterprise partnerships and CPPs, as well as voluntary and community organisations. We now briefly describe each challenge and propose opportunities for policy interventions to address them.

Challenge 1: Many rural residents are at risk of poverty, not a small minority

Half of all rural residents in Britain fell below the poverty line at some time during 1991–2008 (Vera-Toscano et al, 2020), and the FCA (2018) recently found that more than half of all rural residents exhibit financial vulnerability. While a substantial proportion of rural residents are therefore at risk of poverty and experience financial vulnerability, these same studies also reveal that some groups are more at risk, such as older people, young people, lone parents and those who are unemployed. Our research suggests that, due to the pandemic, many more rural residents will be at risk of financial hardship in the near future. In short, many rural residents are at risk of poverty: this is not a situation facing a small minority in 'pockets of rural deprivation'.

While it may still be appropriate to have specific anti-poverty initiatives targeted at particular groups at particular times or in particular places, the welfare system needs to be open to all and to provide a 'safety net' for anyone who finds themselves in poverty for whatever reason, including those who live in rural areas. As DEFRA's Minister for Rural Affairs has acknowledged, no one should be disadvantaged by where they live.

Unfortunately, poverty is frequently perceived to be an urban challenge because this is where it is most visible; but part of the challenge of addressing rural poverty is that it is hidden. This invisibility might be addressed by ensuring that national policy development is informed by local knowledge and local practice from rural areas, not only through the involvement of local stakeholders in more effective rural proofing as policies are being developed and piloted, but also through mechanisms within national government for continuous learning and policy refinement from local experience. As part of such an approach, piloting of new national initiatives should routinely take place in rural localities, and evidence of learning from these attached to policy announcements in the form of rural impact statements. Such an approach is being taken in Islands Communities Impact Assessments in Scotland.[1]

Our research revealed the vital role of a wide range of formal and informal groups in providing support of different kinds at different times. These groups, including voluntary and community organisations, give people different 'entry' points to the welfare system, depending on their individual networks. It is increasingly important that service providers in rural areas play a signposting role, connecting their clients with information and advice. While they may not be experts on every service, they should at least be able to provide baseline information and know where people can find more help.

Challenge 2: There is an imminent rural cost of living crisis

Many rural households already experiencing poverty and financial vulnerability will be hard hit by the substantial increases in energy costs occurring in 2022, and rising inflation. Our research has revealed that many rural dwellers face fuel poverty, higher costs of living, insecure employment and lack of access to services as these are centralised and digitalised.

Fuel poverty is particularly prevalent in rural areas because many properties are not connected to mains gas, therefore having to rely on more expensive and less regulated sources such as oil and LPG or less efficient electric heating systems. Houses tend to be older and poorly insulated, and difficult and costly to retrofit with insulation (TIG, 2022). Furthermore, poor households in rural areas are more likely to live in private rented houses than in urban areas where social housing is more available. On top of this, rural households incur higher transport costs in travelling to access services and employment, often with no public transport available. A recent report by the Scottish government (2021a) notes that a third of households in remote rural areas were in extreme fuel poverty in 2019 compared with 11 per cent of households in the rest of Scotland.

In 2022 and beyond, high rates of inflation are anticipated by the Office for Budget Responsibility, and it is predicted that increases in energy costs and rising inflation will hit rural areas hardest, with 57 per cent of households in the Western Isles and 47 per cent in the Highlands of Scotland experiencing fuel poverty, according to modelling by Energy Action Scotland (2022). In England, about a third of households are predicted to experience fuel poverty in rural west Norfolk, north-east Lincolnshire, Herefordshire and Shropshire, as well as in the Prime Minister's own constituency of Richmondshire (End Fuel Poverty Coalition, 2022).

A comprehensive study by Robinson and Mattioli (2020) has recently estimated and mapped the potential vulnerability of households in Britain to the combination of increased domestic energy costs and rising petrol prices. Their results for England, shown in Figure 8.2, indicate that it is rural households which exhibit the greatest 'double energy vulnerability' and are therefore likely to suffer the greatest financial pressure in the months ahead.

Our interviews reveal the human impact of fuel poverty in rural Britain. One person told us how he had not realised that his retired parents were suffering fuel poverty until he visited them unexpectedly one day and found them sitting in the cold with their coats on and duvets over them – he discovered then that they could only afford to switch the heating on when he was due to visit. Another Perthshire resident told us how they kept the central heating switched off because of the expense, relying instead on collecting free firewood for a wood-burning stove. In Northumberland we learned how some households run out of oil early in the winter and cannot afford to buy more. Despite these challenges, it was encouraging to learn about several organisations/initiatives working on this issue in each place we studied.

Addressing the cost of living crisis and rising fuel poverty facing so many rural households requires action on many levels. Most immediately, it requires benefit levels to rise in line with inflation and for initiatives to promote take-up in rural areas by those who are eligible. Energy-related initiatives such as home insulation, help with the costs of oil buying and energy efficiency measures and advice should be expanded, building on the success of existing local schemes. These could be assisted by grants for the installation of more energy-efficient heating systems for those in most need and by rural proofing and island proofing of new regulations which threaten the viability of rural insulation schemes (TIG, 2022).

For the medium term, encouragement for the generation of renewable energy in rural areas could be given by restoring the incentives withdrawn by the UK government in recent years, investing in the infrastructure necessary to feed increased power from rural areas into the grid and enabling residents of these areas to benefit from the cheaper energy produced there.

Figure 8.2: Areas in England with high likelihood of domestic and transport-related energy poverty (double energy vulnerability)

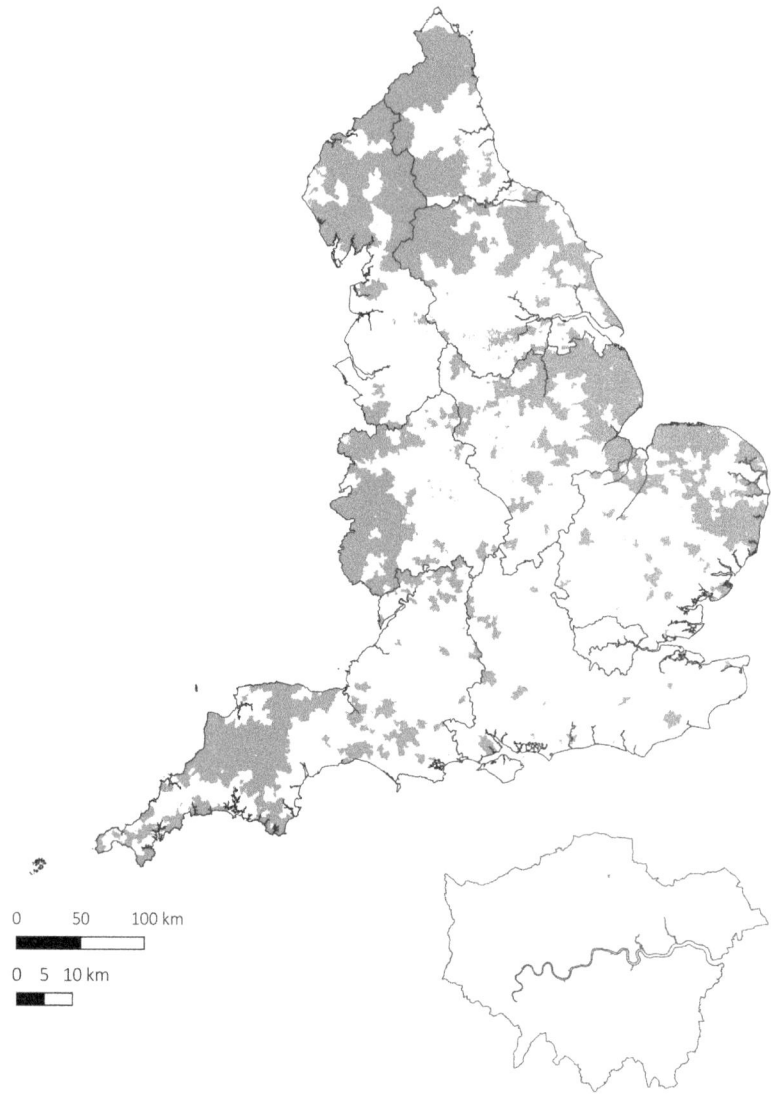

Note: The shading denotes areas with high likelihood of double energy vulnerability.

Source: Robinson and Mattioli (2020). Reprinted from Energy Research & Social Science, Vol 70 (2020), Caitlin Robinson and Giulio Mattioli, 'Double energy vulnerability: spatial intersections of domestic and transport energy poverty in England', Copyright (2020), with permission from Elsevier.

As one islander put it, "the electricity grid was built on a system of power generation near major cities which branched out to supply peripheries. That system now needs to be reversed so that the peripheries can feed renewable energy back to the urban centres". In the longer term, investment should be made in developing local energy grids, to enable the energy generated through renewables such as wind power in rural communities, to be used in the local area.[2]

Challenge 3: The welfare system is not well adapted to rural lives

Some of the challenges highlighted by our research in relation to the welfare system are 'place neutral' in the sense of applying in both rural and urban locations. However, many are particularly rural challenges, evidenced, for example, by lower claimant rates in rural areas. These include distances required to travel for medical assessments, longer waiting times for benefits which require an assessment for work capability nearer to home, the lack of face-to-face information and advice provision locally and feelings of stigma which are amplified in rural locations because of greater visibility.

It is vital, therefore, that DWP should implement rural proofing of the welfare system to address the issues highlighted in this book, with the support of officials from DEFRA, Scottish Government and local government, and with input from relevant stakeholders. In September 2002 DWP's Deputy Director of Work and Welfare Strategy initiated an awayday at Stoke Rochford Hall for senior DWP staff to discuss 'providing DWP services to rural communities', which was perceived by staff as very valuable.[3] This Rural Symposium concluded that DWP needs to consider the impact of policy upon people who are socially excluded and disadvantaged living in rural areas, and that policy and programmes need to be flexible to be effective in helping disadvantaged people in rural communities. A similar symposium would be equally valuable 20 years later.

A priority today, given the digitalisation of services over those 20 years, is for DWP and other service providers to ensure that, where possible, face-to-face provision of welfare advice and support is available, alongside digital or phone provision, particularly for those who may be hard to reach due to physical isolation or digital, literacy or mental health challenges. This may require advice and support organisations to be better funded to cover the additional costs of providing face-to-face support in rural areas at the local level (rather than focused in one main centre), and as a vital component of the social infrastructure in rural areas offering a gateway to state support.

Gathering and sharing information on the ways in which the welfare system is not 'fit for purpose' for rural residents is an essential part of this rural proofing process.

Challenge 4: Much rural work is not 'good work': in particular, incomes tend to be volatile and irregular

Both the UK and Scottish governments have made commitments to 'Good Work' or 'Fair Work',[4] embracing pay and conditions of employment. Our evidence repeatedly illustrates the flexible and irregular nature of the rural labour market and associated individual/household incomes. As a result, welfare benefit claimants in rural areas experience unnecessary payment delays and/or overpayments with clawback because the online claim system cannot deal with such volatility/irregularity in real time. This often leads unnecessarily to hardship and debt.

It can be challenging for some individuals (particularly working-age women with children) to engage in the flexible rural labour market if working hours do not complement caring responsibilities. It is evident from our research that there is a lack of flexible and affordable childcare provision for families in rural areas. Addressing this may require additional financial support for private and voluntary sector childcare providers. There was a general concern in our case studies about the lack of household budgeting skills in vulnerable rural households, an issue which is compounded in a context where incomes are unpredictable, and the cost of living is higher. The task of managing low and/or volatile incomes would be helped by training and support on household budgeting. This might be achieved by resourcing existing advice and financial service providers (CABs, credit unions, money advice services and so on) accordingly. Early intervention in schools might also be helpful in embedding these skills from an early age. To progress towards realising Good/Fair Work objectives, there is a need to address the challenges evident in rural areas of low and irregular wages, seasonality, insecure employment, lack of childcare, few opportunities for training and/or progression and work/life imbalances.

Policies of central and local government, and private businesses, should ensure that the principles of the 'Good/Fair Work' agendas are applied in rural areas in the same way that they are in urban locations. Rural employers need to be encouraged to sign up to these principles, for example, by offering (and being supported to offer) training, upskilling and career progression opportunities. One approach may be for small businesses to work together to deliver such opportunities.

Diversifying rural economies beyond tourism and land-based work should remain a public policy aim – to generate a mix of good employment and self-employment options in rural areas. To redress the imbalance in the local economy, the focus should not be on limiting tourism, but on increasing the strength of other sectors.

Challenge 5: There are barriers to entering self-employment and developing rural small businesses

Diversifying rural economies requires additional support for those wishing to enter self-employment and/or establish a rural microbusiness. We heard that small amounts of start-up funding (less than £10,000) were unavailable for microbusinesses in rural Scotland, and that many rural businesses did not fulfil the criteria for funding related to COVID-19. We found that self-employment is not always seen as an option by individuals facing financial hardship, and those who have become self-employed through necessity are particularly likely to lack confidence/knowledge about what is involved and to require additional information and support.

There are also limited opportunities to access financial and other support for businesses that are not growth-oriented, ruling out many rural microbusinesses that are sole operators or employ only one or two people. Small businesses established by women to allow flexible working around caring responsibilities (such as cleaning or beautician businesses) appear less likely to be supported by funding and support agencies, even though they are viable. Other barriers include literacy issues and a shortage of technical/digital skills among financially vulnerable residents, which not only affects their ability to consider alternative employment options but also to access crucial online information and services that are not provided in person locally.

An important first step should be to recognise the importance of self-employment and small businesses in the rural labour market through making available tailored advice from peers and financial support for sole traders and rural microbusinesses which are not primarily growth-oriented. A small fund to help with living costs while small businesses get off the ground (similar to that available for social enterprises in Scotland through Firstport) would also be helpful. Start-up funding for young people wishing to be self-employed should be available through schemes such as the Kickstart Scheme and the Young Persons' Guarantee Scheme.[5]

Place-based skills audits may help to identify skills gaps among rural residents in relation to self-employment opportunities in the local labour market – these gaps could then be targeted with rural-proofed interventions, such as early intervention in local schools or local mentoring schemes.

Self-employment/microbusiness start-ups can be limited by a lack of attractive and affordable business premises in suitable rural locations – private and public sector creation of small business units/workshops to suit different types of rural businesses are likely to be well used. Rural community hubs can also be a way of providing a desk without an 'office', an office classically being out of financial reach of most start-ups (see also, for example, the study by Cowie et al, 2013, of rural growth hubs in Northumberland). Hubs also address business and social isolation issues for people who would otherwise be working at home, where they miss out on the informal support from being in a business environment.

Finally, large-scale contracts exclude many small rural enterprises from the opportunity to bid because of limited capacity to deliver at the large scale. Public bodies should be encouraged to break up large contracts into smaller ones to enable more small, local enterprises to bid for work – or be required to issue ten small contracts for every large one, for instance.

Challenge 6: It is challenging for organisations to 'reach' into rural areas

Our research notes the tendency towards centralisation and digitalisation, alongside the continued need for face-to-face access to support for those without digital skills or infrastructure, or with poor literacy or mental health. Voluntary and community organisations are clearly filling a gap where public sector support has been withdrawn. However, these face funding challenges due to increased competition for funding and a reduction in volunteer support in some places.

We have seen several good examples of joint working between public and third-sector outreach services to reach rural residents facing financial hardship and encourage higher levels of benefits take-up. 'Everyday' support delivered in the case studies in public community spaces and from trusted community groups is supplementing other forms of support, including that from family and friends. 'Trusted' places and people within the community who offer support can help to address visibility issues/stigma associated with needing help in rural areas.

One of the most important findings from our research concerns the necessity of ensuring people have a plurality of means of accessing services in and from rural areas – that is, providing a mix of face-to-face outreach, mobile, digital and phone services. There were successful examples of this across our study areas, for instance, outreach in GP surgeries, lunch clubs, Warm Hubs and libraries. Moreover, the experience of delivering these services during the pandemic showed that staff could be remotely located rather than based in a central office, and this opens up opportunities for more dispersed local delivery. Of course, reinstating/supporting mobile and outreach face-to-face services for the most vulnerable groups is likely to require additional financial support where delivery of services is not financially viable, yet much needed by rural residents. One way of resourcing and promoting more effective delivery might be through incentivising joint working and making use of existing community assets for local delivery.

There is also scope for many organisations in rural Britain to work with partners to extend public spaces in rural areas in which people without digital access and/or skills can access the internet free of charge on their own device or at a workstation. This should go hand in hand with investing in enhancing social infrastructure (building on the success of the Warm Hubs in Northumbrian village halls and so on) not only to enable social interaction but also to help develop informal and formal networks between agencies, voluntary and community organisations and communities, leading to information sharing, collective insurance, funding, staff sharing and so on.

There is an important role here for community-based networks, local representatives and even researchers to ensure that national or regional organisations understand the particular characteristics of rural areas, so their policies and delivery mechanisms reflect rural contexts, in short, increasing the level of knowledge and intelligence with which to make informed decisions affecting the lives of people living and working in rural areas.

Challenge 7: It remains difficult to access suitable/affordable housing in rural Britain

The evidence gathered in our case studies confirms that long-standing challenges remain in relation to rural housing. It is hard for rural dwellers on irregular or volatile incomes to secure mortgages. The chronic lack of social housing in rural areas presents additional challenges, as do local authority letting policies which can lead to financially vulnerable households being placed in remote locations where there are no job opportunities and no public and/or private transport. Residential housing stock is often not appropriate for the needs of young or older people and single-person households, and there are issues with the quality and cost of private rented accommodation/ tied housing. Older housing stock tends to be associated with poor energy efficiency, which amplifies fuel poverty issues in rural areas. Properties may also be empty with potential for bringing them back into use.

The dominant tourism and hospitality industry brings more demand for second/holiday homes and self-catering accommodation, which reduces the range of properties on the market. This presents an additional issue for tourism businesses who cannot find suitable accommodation for their (often seasonal and low-paid) staff, as well as frustrating young people's ability to stay in the area.

Several recent reports have documented the rural housing challenge, and each of these has proposed a number of potentially effective policy opportunities, often commanding support from rural stakeholders.[6] These include an urgent need for much greater public investment in social rented housing in villages across rural Britain, with appropriate allocation policies to support rural development objectives. Resources and political will are required to implement these.

Although rural housing is recognised as a long-standing policy challenge, there is an opportunity to evidence the need for rural housing in more detail – this requires a focus on vulnerable groups, as well as those who are experiencing hidden homelessness. A national policy framework should enable a place-based and locally informed approach to delivering housing in rural areas, based on this evidence of need – in addition, this process can support new or existing rural business development in the right place at the right time, to ensure rural workers have access to suitable accommodation.

As well as other fiscal and regulatory measures currently under consideration to relieve the pressure on rural housing markets in popular areas, consideration might also be given to permitting holiday cabins, or 'huts' based on the Nordic model (Riddoch, 2020), to take some of the pressure off the housing stock while nevertheless encouraging tourism's contribution to the local economy.

In certain areas, making available grant funding for upgrading and refurbishing housing may also be important to encourage empty housing to be put back into use to meet housing need and assist rural development in ways which are appropriate for current living, working and socialising patterns.

Challenge 8: Framing narratives of place and change are important to the wellbeing of rural communities

As noted earlier, place-based narratives of what has been lost from rural communities were present in all our case studies. Equally, though, narratives of optimism and hope were often founded on a sense of place and cultural identity, notably in Harris but also elsewhere.

Experiences of social isolation, poverty of contacts and concern about stigma and visibility when asking for help were intertwined in narratives of change. For several individuals, their current financial situation was often linked (perhaps subconsciously in some instances) to broader social changes that had led to reduced local opportunity structures being available to them. However, there was optimism in Harris, linked to community-owned assets and the future of the area's economy, and in parts of the North Tyne valley where people could envisage a hopeful future and were working together to realise that. In all places, despite challenges, strong place attachment was evident, even among those who were experiencing quite difficult personal situations.

While it is important not to dismiss or diminish the importance of the histories of places, supporting communities to build positive narratives of place and change – from the bottom up – is important to build confidence and to mobilise. Creating a sense of hope around a shared vision, reflecting both continuity and change, is important: but it needs to pay attention to power imbalances and social inclusion, valuing 'local' claims to cultural capital as well as welcoming new ideas. There may be roles for external organisations here, for example, facilitating local conversations or providing small amounts of funding to stimulate community projects which contribute to positive narrative building while reflecting each place's own strengths and identity, learning from the LEADER and other community-based approaches (see Atterton et al, 2020), and perhaps drawing on notions of 'radical kindness' (Ferguson, 2016).

A genuine approach to community empowerment might encourage and support the community ownership of revenue-bearing assets, such as affordable housing or renewable energy. These strengthen community-based organisations and give them greater power and resilience to pursue their own agreed objectives. In Scotland, targeted business support for community organisations is available (through Just Enterprise, a business support contract from Scottish government) to help them generate long-term income – additional support of this nature in other parts of the UK would be beneficial.

Challenge 9: There is an imminent crisis in rural social care delivery

It is widely acknowledged that the national system for providing social care is poorly resourced and requires urgent reform, and indeed successive governments have committed to such reforms. Announcements on the cap on social care costs and the Social Care White Paper in 2021 were published after our research was completed (King's Fund, 2021). The evidence generated from our work highlights several problems related to social care in rural areas: labour shortages, increasing demand, decline of support from friends and family, long travel times for carers between clients with no support for the associated costs and older people living for longer (often alone) in inappropriate housing.

These issues have implications for individuals as well as the organisations responsible for delivering care. When a national policy addressing the crisis in social care is eventually introduced, our findings highlight the specific rural dimensions that will need consideration within the national approach, hopefully through rural proofing the policy as it is developed. This is especially important given the older population of rural areas in Britain.

A strong message emerging from our research is that any proposed new national policy for social care provision should be rural proofed to ensure that it takes account of the additional costs and other specific challenges of providing social care in rural areas, as highlighted in this report. Rural stakeholders should be fully involved in this process. Similarly, the Scottish government's 'Carers' Wellbeing Campaign' initiative[7] should be alert to the additional challenges of providing and supporting care in rural areas. A crucial element of recognising the distinctive rural challenges of providing social care is that local government and voluntary providers should be adequately resourced to meet the extra costs of provision of social care services in rural areas (including longer travel times).

At the same time, there is untapped potential in rural areas to deliver a more personalised and joined-up approach via informal cooperation between health and care workers, learning from more flexible working practices adopted during the COVID-19 pandemic. This is challenging because of the fragmentation of responsibilities and budgets which are an obstacle to integrated service delivery.

Closing reflections on power and governance

Underlying these experiences of social exclusion in rural Britain are important issues of power and governance. It is apparent that there is a tendency for state services (like many private services) to become increasingly centralised in towns and cities and ever more distant from rural citizens, unless they can be digitalised and issues of digital exclusion overcome. While voluntary and community organisations also face budgetary pressures to centralise their services, there remains an invaluable social infrastructure of village and community halls, voluntary and community organisations and informal networks, often facilitating the activities of others, such as lunch clubs, Warm Hubs, sports clubs and foodbank and advisory services outreach. These are invaluable to many rural citizens, pursuing social objectives, reducing social isolation through volunteering and helping to build community cohesion, while also to some extent covering for gaps in state provision. But, while such support is widely valued and much needed, the question arises of whether a discourse and practice of self-help masks and facilitates the state's failure to fulfil its duties to its rural citizens in a way which might not be tolerated in urban contexts, as well as widening inequalities between places with more or less capacity to act. As Shucksmith et al (2021, 257) put it:

> Should we celebrate the contributions of civil society in mitigating the impact of rolling back the state, or lament civil society's role in masking the state's abdication of its role in serving its citizens? Should we embrace the activities of civil society as resistance to austerity and neoliberalism, or criticise civil society for enabling and facilitating these?

Both viewpoints may be valid, of course: applauding the contributions of voluntary and community organisations and civil society more broadly need not preclude an analysis which identifies a framing narrative of rural community self-reliance and self-help as operating in the service of neoliberalisation (Cheshire, 2006), an example of what Pierre Bourdieu termed 'symbolic violence'. After all, Margaret Thatcher argued in 1979 that the building of the British welfare state after 1945 had 'crowded out' charitable provision[8] and intended her roll-back neoliberalisation to redress this (Fougère et al, 2017, 823, quoted by Curry at the launch of Curry,

2021). An alternative perspective sees charity as anathema (Fougère et al, 2017), recalling distinctions between a 'deserving poor' and the 'undeserving' poor, or as 'roll-with-it' neoliberalisation (Keil, 2009, 232).

An important task of social scientists is to reveal hidden power relations, in the hope that making the exercise of power visible may call its legitimacy into question. But, after revealing the hidden exercise of power and mindful of neoliberalisation, perhaps the crucial question going forwards is how to redesign the relationship between civil society and local state (and implicitly central state) after austerity and after the COVID-19 pandemic. How might a productive relationship be framed with contributions from both? And how can this be translated from local experience to national policy?

This is more than just a question of how to work well in partnerships, about which so much has been written (for example, Westholm et al, 1999; Moseley, 2003). At its heart, it is about how the state can fulfil the roles of provider, regulator and enabler (Healey, 2006; Shucksmith, 2010) exercising not only 'power over' but also building its citizens' 'power to'; it is about the potential for synergy between representative and participatory forms of democracy (see Commission for Rural Communities, 2008); and it is about community empowerment, localism, social innovation and social inclusion. How far does the asymmetrical power relationship between the state and voluntary and community organisations limit the practical potential for localised and joined-up action, for example? Exploring these questions fully would occupy another book, but it is apparent from previous studies that there is considerable diversity across Britain in how willing local councils have been to work with voluntary and community organisations and that there is much scope for sharing learning. In this study we found many examples of creative joint working in operations and delivery but heard voluntary and community organisations express frustration that they were often excluded from strategic processes. We learned that many Community Development Trusts, for example in Tynedale, had received crucial support from council officers in their early establishment and development, and of the vital role of the Highlands and Islands Enterprise Community Land Unit and the Scottish Land Fund in the realisation of Scotland's community land movement (see also Glass et al, 2021).

Another element is the concentration of power in the UK central state (Westminster and Whitehall), the ongoing disempowerment of English local councils and the intensification of central control by Westminster despite a rhetoric of localism. Respondents perceived this both in terms of urban power being exercised over rural citizens and as centralised Westminster power being exercised from afar over local matters, in each respect without an understanding of the local/rural context and without listening to local stakeholders or respecting their local knowledge and legitimacy. 'Policies that position cities as drivers of economic growth, for example, can condemn rural communities to chronically inadequate infrastructures while idealistically requisitioning "unspoilt" landscapes for urban leisure consumption" (Woods et al, 2019). In the Scottish study areas, there were also tensions over devolved and reserved powers, with Scottish government policies (for example, on social security or payment of Housing Benefit direct to landlords) widely viewed as more caring and kinder than those of the UK government. Moreover, a national outcome agreement between councils and the Scottish government contrasted with the deep cuts in funding for councils and the centralised exercise of power in/from England.

These perceptions of rural citizens' disempowerment are not unique to rural Britain: 'the perceived marginalisation of rural communities in conflicts over rural space has contributed to an increasing sense of beleaguerment and neglect by rural communities in many parts of the global north', giving rise to 'a new rural identity movement' which is 'distinctive in its articulation of a rural–urban divide as a framing device' (Woods 2011, 287). However, in Britain (and especially in England) rural citizens face one of the world's most highly centralised, top-down governance structures alongside growing spatial inequalities (McCann, 2016), so lending force to rural citizens' feelings of neglect. Post-Brexit rural policies (House of Lords, 2019; Shucksmith, 2019) remain to be developed and, notwithstanding the UK government's recent White Paper on 'levelling up', detailed proposals for addressing spatial inequalities are still awaited and rural stakeholders are sceptical that any attention will be paid to the distinctive and diverse circumstances of Britain's rural areas. Thus, the Rural Coalition (2022, 1) were:

> disappointed that, despite a few references to rural areas/communities/villages, the tenor of the White Paper was essentially urban and city focussed. The overriding philosophy presented for levelling up appears to be based on the assumption that growth and improvement derives primarily from cities and urban aggregation and that success in these will trickle down to surrounding areas.

They call for the government now to set out its promised vision for rural England and develop a cross-cutting strategy for how levelling up can be achieved for rural people, businesses and places. Speaking directly to the concept of rural citizenship, the Rural Coalition concludes that levelling up must deliver on the philosophy that 'no-one should be left behind due to where they live'.

The House of Lords Select Committee on the Rural Economy recommended the development of a rural strategy, improved rural proofing and a place-based approach reflecting this diversity:

> The rural areas of England vary enormously, as do the economies within them. Any rural strategy and the policies that flow from it must

take these variations into account and ensure that local communities are fully engaged. We recommend that the national rural strategy enables, and is realised through, a 'place-based approach', meaning one that is connected to local needs and interests, and with the participation of as wide a range as possible of public and private bodies, community groups, businesses and individuals. (House of Lords, 2019, 9)

That Committee also argued that nobody should be disadvantaged by where they live, and indeed DEFRA ministers have spoken of their determination to ensure 'people living in our market towns and villages have the same life opportunities as those who live in our cities' – something which is a statutory right in Norway (Shortall and Alston, 2016) but is lacking in rural Britain (Social Mobility Commission, 2019). Rural citizens (in the broadest sense) should expect fairness and similar rights of citizenship – that is, fair outcomes including access to services which meet needs, investment in social and economic infrastructure, transparent decisions based on evidence, equal opportunities to participate in society and a fair hearing and an effective voice in decision making (Shucksmith, 2019).

The neglect of poverty and social exclusion in rural Britain by policy and practice arises both from a lack of understanding of these circumstances by those in power but more fundamentally from the exercise of power to further other, more dominant interests. Cheshire (2016, 593) has detailed how the exercise of power operates within and across rural areas 'to serve the interests of some, but not others, and to generate marginalisation, inequality, conflict, resistance and change'. Power is exercised not only through formal decision-making processes and less visible agenda setting, but also through a more insidious 'third face of power' (Lukes, 1974) which shapes people's perception of what rurality is and what rural areas should be. Lukes showed how power could be exercised over people 'by shaping their perceptions, cognitions and preferences in such a way that they accept their role in the existing order of things' (Lukes, 2005, 28), because they can see no alternative, or they see it as natural and unchangeable. Inevitably, rural poverty and social exclusion are exacerbated by framings of poverty and social exclusion as urban, not rural, issues, or by framings of rural communities' greater social cohesion and potential for self-help absolving the state from its responsibility to rural citizens.

Returning to the four sources of support in our analytical framework (markets, state, voluntary and community sector, and family and friends), we can now summarise how they relate to social exclusion in rural Britain. Much rural employment is precarious, low paid or seasonal, with volatile and unpredictable incomes creating financial vulnerability. For households in such circumstances, a series of welfare reforms has reduced the real value of benefits, while intensifying work activation and conditionality. Specific rural dimensions to this arise from the volatility of rural incomes, digitalisation and digital exclusion, difficulties in accessing advice and support and typically lower claimant rates. As a further layer on top of this, people experience higher costs of living and widespread fuel poverty, portrayed by some as constituting 'double disadvantage'. Meanwhile, voluntary and community organisations have been active in supporting many of the people disadvantaged by markets and the state, and their services are highly valued, but they face challenges of reach into rural areas as well as resource constraints which tend towards further centralisation and digitalisation. Family, friends and neighbours are another important source of support, but people's ability to draw on this source varies according to the characteristics of the community they live in, its social norms, social capital and personal relationships. There is a tendency to idealise rural communities as places where everyone looks after one another, and there is a concern that this reputation for self-help and social cohesion facilitates and legitimises a withdrawal of state welfare services from rural areas. In this chapter we have reflected on how policy and practice might respond to these challenges: the first step is to acknowledge that rural poverty and social exclusion exist and that policies and practice will be more effective if there is a better understanding of experiences in their local contexts.

Notes

Chapter 1

[1] Poor households were defined by McLaughlin as being those whose incomes were below the then welfare system's supplementary benefit threshold, and households on the margin of poverty were defined as those with an income up to 139 per cent of this threshold.

Chapter 2

[1] ACORN segmentation is a widely used consumer classification technique, developed by CACI, which analyses demographic data, social factors, population and consumer behaviour to classify areas. Further details are available at: https://acorn.caci.co.uk/

[2] Wisecraft is part of the Perth and Kinross Association of Voluntary Service Ltd (PKAVS) Mental Health and Wellbeing Hub, based in Blairgowrie. It provides a safe environment for adults recovering from mental illness to rebuild their lives. It offers meaningful work-related experience, stimulates creativity through arts and crafts and builds confidence and self-esteem, as well as offering help and support. See: www.pkavs.org.uk/Wisecraft

[3] All council tenants were given the right to buy their council houses at heavily discounted prices by the Conservative government in the Housing Act 1980. Council house building ceased as a result, and over half of the pre-existing council houses were purchased and are no longer available to those on waiting lists.

Chapter 3

[1] A croft is a small agricultural unit that is usually rented, and is a form of tenure under Scots law. Crofts are often part of large estates where the landowner is the crofter's landlord. The croft often comes with the right to graze animals like sheep or cows on the common grazings, an area of land shared with other crofts. Crofts are located in one of the crofting counties or other areas designated by the Scottish government.

[2] Harris Development Limited is a not-for-profit local development company that provides a coordinating role for third-sector organisations across Harris, to drive community-led development. See: www.harrisdevelopment.co.uk/

[3] Angus MacNeil MP wrote to Atos Healthcare in September 2013. More information is available on his website: https://angusmacneilsnp.com/2013/09/26/macneil-anger-at-atos-healthcare-constituent-lost-in-system/

[4] TIG is a Registered Society under the Co-operative and Community Benefit Societies Act 2014 that is run primarily for the benefit of the Outer Hebrides community. It supports people to rent, buy and live in comfortable, affordable homes, promotes healthy independent living and assists communities and businesses to be more sustainable. Specifically, TIG delivers support to those disadvantaged by age, income or disability to insulate, heat and maintain their homes through a variety of advice and funding schemes.

[5] The RET is a distance-based fares structure, which underpins the Scottish government's commitment to providing one single overarching fares policy across Scotland's entire ferry network. The formula for calculating ferry fares is a combination of a fixed element and a rate per mile. See: www.transport.gov.scot/public-transport/ferries/road-equivalent-tariff/

[6] The Place Standard is a tool that can be used to assess the quality of a place. It can assess places that are well established, undergoing change or still being planned. The tool can also help people to identify their priorities for a particular place. See: www.placestandard.scot/

Chapter 5

[1] Research published in May 2021 by Hirsch and Stone (2021) found that levels of child poverty in Northumberland increased from 26.5 per cent in 2014–15 to 36.2 per cent in 2019–20. This is an increase of 9.8 percentage points and the increase in Northumberland is ranked 12th highest when compared against all UK local authority areas. See: www.nechildpoverty.org.uk/facts/

[2] These data were collected from the postcode checker available at: http://inorthumberland.org.uk/

[3] Tarset's Submission Draft Plan (June 2015) is available on the community website: www.tarset.co.uk/community/submission-draft.cfm

[4] More information about the Community Action Northumberland 'Warm Hubs' can be found on the CAN website here: http://ca-north.org.uk/supporting-individuals/warm-hubs

[5] The Employment Hubs are part of the wider 'Bridge Northumberland' project: www.bridgenorthumberland.org.uk/

Chapter 6

[1] This is also the case in other rural regions – see Highlands and Islands Enterprise (2020).

[2] More information about the work of Northumberland Communities Together is available on the Northumberland County Council website: www.northumberland.gov.uk/coronavirus/Northumberland-Communities-Together.aspx

[3] See: https://ihub.scot/media/8233/community-health-and-wellbeing_final-report.pdf

Chapter 8

[1] Islands Communities Impact Assessments are a provision of The Islands (Scotland) Act 2018 and require relevant authorities to consider formally how policy, strategy and/or services impact on island communities. See Atterton (2019) for more information.

[2] See www.theguardian.com/environment/2015/feb/12/renewable-energy-wind-changed-fortunes-lewis-islanders.

[3] DWP Rural Symposium, 'Providing DWP services to rural communities', Stoke Rochford Hall, 10 September 2002. Official minutes, 17 pages.

[4] See the UK government Good Work Plan here: www.gov.uk/government/publications/good-work-plan, and the Scottish government Fair Work Action Plan here: https://economicactionplan.mygov.scot/fair-work/

[5] More information about each of these schemes is available. For Firstport see www.firstport.org.uk/funding/; for the Kickstart Scheme see: www.gov.uk/government/collections/kickstart-scheme; and for the Young Persons' Guarantee see www.myworldofwork.co.uk/youngpersonsguarantee

[6] For example, the 'Rural housing policy review for the National Association of Local Councils' (2015), available at: www.nalc.gov.uk/nalc-blog/entry/190-rural-housing-policy-review; 'Rural recovery and revitalisation report to the Rural Services Network' (2020), available at: www.rsnonline.org.uk/publications; and 'The role of land in enabling new housing supply in rural Scotland – report to the Scottish Land Commission' (2020), available at: https://bit.ly/3aoJQua

[7] See: www.nhsinform.scot/caring

[8] 'As Thatcher herself once put it (quoted by Mirowski, 2013, 95), "faith-based charities … were crowded out by the rise of the welfare state and would grow again … if government were to reduce its profile or remove itself entirely"' (Fougère et al, 2017, 823). Thanks to Nigel Curry for this reference.

References

ACRE (Action with Communities in Rural England) (2022) 'Is the government doing enough to stave off the cost-of-living crisis in rural areas?' [blog], available from: https://acre.org.uk/is-the-government-doing-enough-to-stave-off-the-cost-of-living-crisis-in-rural-areas/ [accessed 26 May 2022].

Arshed, N. (2021) 'The impact of COVID-19 on Scotland's women entrepreneurs', Scottish Parliament Information Service (SPICe) briefing, SB 21–73.

Asenova, D., McKendrick, J., McCann, C. and Reynolds, R. (2015) 'Redistribution of social and societal risk: the impact on individuals, their networks and communities in Scotland', York: Joseph Rowntree Foundation.

Atterton, J. (2008) 'Rural proofing in England: a formal commitment in need of review', Centre for Rural Economy Discussion Paper Series No. 20.

Atterton, J. (2019) 'Learning lessons from early Islands Communities Impact Assessments', report for Scottish Government, Rural Policy Centre, Edinburgh: Scotland's Rural College.

Atterton, J., Meador, E., Markantoni, M., Thomson, S. and Jones, S. (2019) 'Exploring the gender pay gap in rural Scotland', report for Scottish Government, Rural Policy Centre, Edinburgh: Scotland's Rural College.

Atterton, J., McMorran, R., Glass, J., Jones, S. and Meador, E. (2020) 'The role of the LEADER approach post-Brexit', report for Scottish Government, Rural Policy Centre, Edinburgh: Scotland's Rural College.

Audit Scotland (2021) 'Local government in Scotland: financial overview 2019/20', Edinburgh: Accounts Commission.

Bailey, N., Spratt, J., Pickering, J., Goodlad, R. and Shucksmith, M. (2004) 'Deprivation and social exclusion in Argyll and Bute', report to the Community Planning Partnership by the Scottish Centre for Research on Social Justice, Glasgow: University of Glasgow, with the Arkleton Centre for Rural Development Research, Aberdeen: University of Aberdeen.

Bailey, N., Bramley, G. and Gannon, M. (2016) 'Poverty and social exclusion in urban and rural areas of Scotland', Working Paper, Poverty and Social Exclusion, Bristol: PSE UK.

Beck, D., Land, E., Gwilym, H., Harris, I. and Gwilym, H. (2016) 'Mapping the growth of the Welsh food bank landscape 1998–2015', Bangor: Bangor University.

Beck, U. (1992) *Risk Society: Towards a New Modernity*, London: Sage.

Beck, U. (2000) 'Living your own life in a runaway world: individualisation, globalisation and politics', in W. Hutton and A. Giddens (eds) *On the Edge: Living with Global Capitalism*, London: Jonathan Cape, pp 164–74.

Bernard, J. (2019) 'Families and local opportunities in rural peripheries: intersections between resources, ambitions and the residential environment', *Journal of Rural Studies*, 66: 43–51.

Bernard J., Contzen S., Decker A. and Shucksmith, M. (2019) 'Poverty and social exclusion in diversified rural contexts', *Sociologia Ruralis*, 59(3): 353–68.

Bertolini, P., Montanari, M. and Peragine, V. (2008) 'Poverty and social exclusion in rural areas: final study report', report to the European Commission, Brussels: European Commission.

Black, N., Scott, K. and Shucksmith, M. (2019) 'Social inequalities in rural England: impacts on young people post-2008', *Journal of Rural Studies*, 68: 264–75.

Boyle, S. (2020) 'How Scotland tackles inequality is my top priority' [blog], 16 November, available from: https://auditscotland.wordpress.com/2020/11/16/how-scotland-tackles-inequality-is-my-top-priority/ [accessed 15 January 2022].

Bradley, A. (1987) 'Poverty and dependency in village England', in P. Lowe, A. Bradley and S. Wright (eds) *Deprivation and Welfare in Rural England*, Norwich: Geo Books, pp 151–78.

Bramley, G., Lancaster, S. and Gordon, D. (2000) 'Benefit take-up and the geography of poverty in Scotland', *Regional Studies*, 34(6): 507–19.

Brewer, M. and Handscomb, K. (2020) 'This time is different – Universal Credit's first recession: assessing the welfare system and its effect on living standards during the coronavirus epidemic', London: Resolution Foundation.

Brewer, M. and McCurdy, C. (2021) 'Post-furlough blues. What happened to furloughed workers after the end of the Job Retention Scheme?', London: Resolution Foundation.

Brewer, M., Francesconi, M., Gregg, P. and Grogger, J. (2009) 'In-work benefit reform in a cross-national perspective – introduction', *Economic Journal*, 119 (February): F1–F14.

Brown, D. and Hirschl, T. (1995) 'Household poverty in rural and metropolitan-core areas of the United States', *Rural Sociology*, 60(1): 44–66.

Bryce, R., Johnson, V., Davidson, M., Heddle, D. and Taylor, S. (2021) 'Community change-scapes of COVID-19 recovery – cross-case report for the Highlands and Islands', Inverness: University of the Highlands and Islands.

Bulman, M. (2017) 'Beautiful countryside hides ugly truth of social isolation, poor health and poverty, report says', The Independent, [online] 18 March, available from: www.independent.co.uk/news/uk/home-news/rural-communities-countryside-public-health-england-local-government-association-neglected-digital-a7636521.html [accessed 15 January 2022].

Bynner, C., McBride, M. and Weakley, S. (2021) 'The COVID-19 pandemic: the essential role of the voluntary sector in emergency response and resilience planning', Voluntary Sector Review, 13(1), doi: https://doi.org/10.1332/204080521X16220328777643 16 July.

Cappellari, L. and Jenkins, S.P. (2009) 'The dynamics of social assistance benefit receipt in Britain', ISER Working Paper no. 2009–29, Colchester: Institute for Social and Economic Research, Essex University.

Chapman, P., Phimister, E., Shucksmith, M., Upward, R. and Vera-Toscano, E. (1998) *Poverty and Exclusion in Rural Britain: The Dynamics of Low Income and Employment*, York: York Publishing Services.

Cheshire, L. (2006) *Governing Rural Development: Discourses and Practices of Self-Help in Australian Rural Policy*, Aldershot: Ashgate.

Cheshire, L. (2016) 'Power and governance: empirical questions and theoretical approaches for rural studies', in M. Shucksmith and D. Brown (eds) *Routledge International Handbook of Rural Studies*, London: Routledge, pp 593–600.

Children in Wales (2008) 'Families not areas suffer rural disadvantage: support for rural families in Wales', Cardiff: Children in Wales.

Children's Neighbourhoods Scotland (CNS) (2018) 'Using the capabilities approach with children and young people', Glasgow: CNS.

Citizens Advice Service (2018) 'Bringing food to the table: findings from the Citizens Advice Scotland national survey on food affordability, access and availability', Edinburgh: Citizens Advice Service.

Cloke, P. and Milbourne, P. (1992) 'Deprivation and rural lifestyles in rural Wales: rurality and the cultural dimension', *Journal of Rural Studies*, 8: 360–74.

Cloke, P. and Little, J. (eds) (2005) *Contested Countryside Cultures: Otherness, Marginalisation and Rurality*, London and New York: Routledge.

Cloke, P., Milbourne, P. and Thomas, C. (1994) 'Lifestyles in rural England', number 18, Cheltenham: Rural Development Commission.

Cloke, P., Goodwin, M., Milbourne, P. and Thomas, C. (1995a) 'Deprivation, poverty and marginalisation in rural lifestyles in England and Wales', *Journal of Rural Studies*, 11: 351–65.

Cloke, P., Milbourne, P. and Thomas, C. (1995b) 'Poverty in the countryside', in C. Philo (ed) *Off the Map: The Social Geography of Poverty in the UK*, London: CPAG, pp 83–102.

Cloke, P., Milbourne, P. and Thomas, C. (1997) 'Living lives in different ways: deprivation, marginalisation and changing lifestyles in rural England', *Transactions of the Institute of British Geographers*, 22: 210–30.

Cloke, P., Widdowfield, R. and Milbourne, P. (2000a) 'The hidden and emerging spaces of rural homelessness', *Environment and Planning A*, 32(1): 77–90.

Cloke, P., Milbourne, P. and Widdowfield, R. (2000b) 'Homelessness and rurality', *Society and Space*, 8: 715–35.

Cloke, P., Johnsen, S. and May, J. (2007) 'The periphery of care', *Journal of Rural Studies*, 23: 387–401.

Combe, M., Glass, J. and Tindley, A. (2020) *Land Reform in Scotland: history, law and policy*, Edinburgh: Edinburgh University Press.

Comhairle nan Eilean Siar (2019) 'Socio-economic update no 41, December 2019', Stornoway: Comhairle nan Eilean Siar Communities Department.

Comhairle nan Eilean Siar (2020) 'Outer Hebrides local child poverty action report', Stornoway: Comhairle nan Eilean Siar.

Commins, P. (ed) (1993) 'Combatting exclusion in Ireland, 1990–94: a midway report', Brussels: European Commission.

Commission of the European Communities (1993) 'Background report: social exclusion – poverty and other social problems in the European Community', Luxembourg: European Commission.

Commission for Rural Communities (2006a) 'Rural disadvantage: quality of life and disadvantage among older people – a pilot study', written by Thomas Scharf and Bernadette Bartlam, Cheltenham: Commission for Rural Communities.

Commission for Rural Communities (2006b) 'Rural disadvantage: priorities for action', Cheltenham: Commission for Rural Communities.

Commission for Rural Communities (2007a) 'Pension Credit take-up in rural areas – state of the countryside update', SOC update report 4, Cheltenham: Commission for Rural Communities.

Commission for Rural Communities (2007b) 'Working age benefit claimants in rural England 2000–2006 – state of the countryside update', Cheltenham: Commission for Rural Communities.

Commission for Rural Communities (2008) 'Participation inquiry: strengthening the role of local councillors', Cheltenham: Commission for Rural Communities.

Corfe, S. (2018) 'What are the barriers to eating healthily in the UK?', London: Social Market Foundation.

Couch, D.L., O'Sullivan, B. and Malatzky, C. (2020) 'What COVID-19 could mean for the future of "work from home": the provocations of three women in the academy', *Gender Work Organisation*, 28(S1): 266–75.

Coutts, P., Ormston, H., Pennycock, L. and Thurman, B. (2020) 'Pooling together: how community hubs have responded to the COVID-19 emergency', Dunfermline: Carnegie UK.

Cowie, P., Thompson, N. and Rowe, F. (2013) 'Honey pots and hives: maximising the potential of rural enterprise hubs' Centre for Rural Economy Research Project, Newcastle: Newcastle University.

Cresswell, T. (2009) 'The prosthetic citizen: new geographies of citizenship,' *Political Power and Social Theory*, 20: 259–73.

Crisp, R., McCarthy, L., Parr, S. and Pearson, S. (2016) 'Community-led approaches to reducing poverty in neighbourhoods: a review of evidence and practices', York: Joseph Rowntree Foundation.

Currie, M., McMorran, R., Hopkins, J., McKee, A., Meador, E., Wilson, R., Glass, J., et al (2021) 'Understanding the response to COVID-19: exploring options for a resilient social and economic recovery in Scotland's rural and island communities', Edinburgh: Scottish Government.

Curry, N. (2021) *Reaping a Community Harvest*, Cheltenham: CCPR.

Curtin, C., Haase, T. and Tovey, H. (1996) *Rural Poverty in Ireland*, Dublin: Oak Tree Press.

Davis, A., Hill, K., Hirsch, D. and Padley, M. (2016) 'A minimum income standard for the UK in 2016', York: Joseph Rowntree Foundation.

Davis, A., Hirsch, D., Padley, M. and Shepherd, C. (2021) 'A minimum income standard for the UK in 2021 (summary)', York: Joseph Rowntree Foundation.

Dax, T. and Machold, I. (eds) (2002) *Voices of Rural Youth: A Break with Traditional Patterns?*, Vienna: Bundesanstalt für Bergbauernfragen.

DEFRA (2020) 'Statistical digest of rural England', Bristol: Department of Environment, Food and Rural Affairs.

DEFRA (2021) 'Statistical digest of rural England', Bristol: Department of Environment, Food and Rural Affairs.

Dunn, M.L. (2011) 'Affordable housing in Northumberland National Park', PhD thesis, Newcastle upon Tyne: University of Northumbria.

End Fuel Poverty Coalition (2022) 'Revealed: the streets where nearly everyone is in fuel poverty' [online], available from: www.endfuelpoverty.org.uk/revealed-the-streets-where-nearly-everyone-is-in-fuel-poverty/ [accessed 26 May 2022].

Energy Action Scotland (2022) 'Fuel poverty set to break the 50% barrier in parts of Scotland' [EAS News Release], available from: https://www.endfuelpoverty.org.uk/revealed-the-streets-where-nearly-everyone-is-in-fuel-poverty/ [accessed 29th April 2022].

ERS Research & Consultancy (2021) 'Community needs in the Tyne Valley: The Squires Foundation', Newcastle upon Tyne: The Squires Foundation.

European Network for Rural Development (2017) 'A focus on rural proofing', *ENRD Magazine*, Autumn/Winter 2017: 28–35.

FCA (2018) 'The financial lives of consumers across the UK', London: Financial Conduct Authority.

Ferguson, Z. (2016) 'Kinder communities: the power of everyday relationships', Dunfermline: Carnegie Trust UK.

Fernández-Reino, M. and Rienzo, C. (2022) 'Migrants in the UK labour market: an overview', The Migration Observatory Briefing, Oxford: University of Oxford.

Fischer, A. and McKee, A. (2017) 'A question of capacities: community resilience and empowerment between assets, abilities and relationships', *Journal of Rural Studies*, 54: 187–97.

Fisher, T. (2021) 'NPA COVID-19 response project on economic impacts – main report. Part 1 key findings, recommendations and summary', Stockholm: Nordregio.

Food Research Council (2014) 'Square meal: why we need a new recipe for farming, wildlife, food and public health', Food Research Collaboration, London: Centre for Food Policy, City University London..

Fougère, M., Segerkrantz, B. and Seeck, H. (2017) 'A critical reading of the EU's social innovation policy discourse: (re)legitimizing neoliberalism', *Organization*, 24(6): 819–43.

Francis-Devine, B. (2021) 'Poverty in the UK', House of Commons Library Briefing Paper 7096.

Generation Scotland (2021) 'Rural Covid life survey general report', Edinburgh: Generation Scotland.

Glass, J. and Meador, E. (2020) 'On the margins? Understanding financial hardship in rural areas: working paper 2 – context analysis', Edinburgh: Standard Life Foundation.

Glass, J., Bynner, C. and Chapman, C. (2020) 'Children and young people and rural poverty and social exclusion: a review of evidence', Glasgow: Children's Neighbourhoods Scotland.

Glass, J., Atterton, J., Maynard, C.M., Craigie, M.C., Jones, S.J., Currie, M., Pinker, A. and McKee, A. (2021) 'Facilitating local resilience: case studies of place-based approaches in rural Scotland', Edinburgh: Scottish Government.

Gloyer, M. (2002) 'Living in the countryside: rural disadvantage and rural regeneration', report number 45, Manchester: Centre for Local Economic Strategies.

Halfacree, K. (1993) 'Locality and social representation: space, discourse and alternative representations of the rural', *Journal of Rural Studies*, 9(1): 23–37.

Halfacree, K. (2007) 'Trial by space for a "radical rural" introducing alternative localities, representations and lives', *Journal of Rural Studies*, 23: 125–41.

Hamilton, R. (2021) 'Inclusive economy board: economic update and CSR overview', presentation to the North of Tyne Combined Authority, 25 November.

Hastings, A., Bailey, N., Gannon, M., Besemer, K. and Bramley, G. (2015) 'Coping with the cuts? The management of the worst financial settlement in living memory', *Local Government Studies*, 41(4): 601–21.

Hastings, A., Bailey, N., Bramley, G. and Gannon, M. (2017) 'Austerity urbanism in England: the regressive redistribution of local government services and the impact on the poor and marginalised', *Environment and Planning A*, 49: 2007–24.

Healey, P. (2006) *Collaborative Planning: Shaping Places in Fragmented Societies*, London: Macmillan.

Healey, P. (2018) 'Creating public value through caring for place', *Policy & Politics*, 46(1): 65–79.

Healey, P. (2022) *Caring for Place: Community Development in Rural England*, London: Routledge.

Highlands and Islands Enterprise (2020) 'The impact of COVID-19 on the Highlands and Islands', Inverness: Highlands and Islands Enterprise.

Hirsch, D. (2013) 'Addressing the poverty premium', York: Joseph Rowntree Foundation.

Hirsch, D. and Stone, J. (2021) 'Local indicators of child poverty after housing costs, 2019/20', Loughborough: Loughborough University.

Hirsch, D., Bryan, A., Davis, A., Smith, N., Ellen, J. and Padley, M. (2013) 'A minimum income standard for remote rural Scotland', Inverness: Highlands and Islands Enterprise.

Hirsch, D., Bryan, A. and Ellen, J. (2016) 'A minimum income standard for remote rural Scotland: a policy update', Inverness: Highlands and Islands Enterprise.

House of Lords (2019) 'Towards a rural economy strategy', final report, Westminster, London: House of Lords Select Committee on Rural Economy.

International Labour Organisation (2020) 'Youth and COVID-19: impacts on jobs, education, rights and mental well-being', Geneva: International Labour Organisation.

Jentsch, B. and Shucksmith, M. (2003) 'Education and individualisation among young people in Angus, Scotland', in T. Dax and I. Machold (eds) *Voices of Rural Youth: A Break with Traditional Patterns?*, Vienna: Bundesanstalt für Bergbauernfragen, pp 38–58.

Jentsch, B. and Shucksmith, M. (2004) *Young People in Rural Areas of Europe*, Aldershot: Ashgate.

Joseph Rowntree Foundation (2018) 'Poverty 2018', York: Joseph Rowntree Foundation.

Joseph Rowntree Foundation (2020) 'Poverty 2019/20 findings', York: Joseph Rowntree Foundation.

Joseph Rowntree Foundation (2021) 'Poverty 2021', York: Joseph Rowntree Foundation.

Joseph Rowntree Foundation (2022) 'UK Poverty 2022: the essential guide to understanding poverty in the UK', York: Joseph Rowntree Foundation.

Keen, R. (2016) 'Welfare savings 2010/11 to 2020/21', House of Commons Library Briefing Paper 7667.

Keil, R. (2009) 'The urban politics of roll-with-it neoliberalisation', *City*, 13(2–3): 230–45.

Kelleher, E. (2021). 'Operational overview of 2020/21', Project Co-ordinator AGM report, 24 August, Perth: Perth and Kinross foodbank.

Kempson, E. and Evans, J. (2020) 'How effective are the coronavirus safety nets? An overview of government support using findings from a national survey', Edinburgh: Standard Life Foundation.

King's Fund (2021) 'The Social Care White Paper: not wrong, just not moving far enough in the right direction', *The King's Fund* [blog], 2 December, available from: www.kingsfund.org.uk/blog/2021/12/social-care-white-paper [accessed 15 January 2022].

Lambie-Mumford, H. and Green, M. (2017) 'Austerity, welfare reform and the rising use of food banks by children in England and Wales', *Area*, 49(3): 273–9.

Lichter, D. and Graefe, R.D. (2011) 'Rural economic restructuring: implications for children, youth, and families', in K.E. Smith and A.R. Tickamyer (eds) *Economic Restructuring and Family Well-Being in Rural America*, University Park: Penn State Press, pp 25–39.

Lindsay, C., McCracken, M. and McQuaid, R. (2003) 'Unemployment duration and employability in remote rural labour markets', *Journal of Rural Studies*, 19(2): 187–200.

Lister, R. (2004) *Poverty*, Cambridge: Polity Press.

Little, J. and Morris, C. (2002) 'The role and contribution of women to rural economies', Cheltenham: Countryside Agency.

Littlejohn, D. (2020) 'The economic impact of COVID-19', presentation to Perth and Kinross Community Planning Partnership Board, 7 July.

Loopstra, R., Reeves, A., Taylor-Robinson, D., Barr, B., McKee, M. and Stuckler, D. (2015) 'Austerity, sanctions and the rise of food banks in the UK', *British Medical Journal*, 350: 1775.

Lowe, P. and Speakman, L. (eds) (2006) *The Ageing Countryside: The Growing Older Population of Rural England*, London: Age Concern England.

Lowe, P., Marsden, T., Murdoch, J. and Ward, N. (2012) *The Differentiated Countryside*, London: Routledge.

Lukes, S. (1974) *Power: A Radical View*, Basingstoke: Palgrave Macmillan.

Lukes, S. (2005) *Power: A Radical View* (2nd edn), Basingstoke: Palgrave Macmillan.

Lupton, D. (2020) 'Doing fieldwork in a pandemic', Keele: Keele University.

Mackenzie, A.F.D. (1998) 'The cheviot, the stag … and the white, white rock?' *Environment & Planning D*, 16(5): 509–32.

Mackenzie, A.F.D. (2013) *Places of Possibility: Property, Nature and Community Land Ownership*, Chichester: Wiley-Blackwell, Antipode Book Series.

Maclaren, A.S. and Philip, L.J. (2021) 'COVID-19 and the countryside. Rural areas: distant from the pandemic?', *Geography Directions* [blog], 3 June, available from: https://blog.geographydirections.com/2021/06/03/covid-19-and-the-countryside-rural-areas-distant-from-the-pandemic/ [accessed 15 January 2022].

May, J., Williams, A., Cloke, P. and Cherry, L. (2020) 'Still bleeding: the variegated geographies of austerity and food banking in rural England and Wales', *Journal of Rural Studies*, 79: 409–24.

McCann, P. (2016) *The UK Regional-National Economic Problem*, London: Routledge.

McKee, K., Hoolachan, J. and Moore, T. (2017) 'The precarity of young people's housing experiences in a rural context', *Scottish Geographical Journal*, 133(2): 115–29.

McKendrick, J.H., Barclay, C., Carr, C., Clark, A., Holles, J., Perring, E. and Stien, L. (2011) 'Our rural numbers are not enough: an independent position statement and recommendations to improve the identification of poverty, income inequality and deprivation in rural Scotland', Edinburgh: Scottish Government.

McLaughlin, B.P. (1986) 'The rhetoric and the reality of rural deprivation', *Journal of Rural Studies*, 2: 291–307.

McLaughlin, B.P. (1991) 'Deprivation – 20 years on', *Rural Viewpoint*, 42: 1–2.

Milbourne, P. (2004) *Rural Poverty: Marginalisation and Exclusion in Britain and the United States*, Abingdon: Routledge.

Milbourne, P. (2006) 'Poverty, social exclusion and welfare in rural Britain', in J. Midgley (ed) *A New Rural Agenda*, London: Institute for Public Policy Research, pp 76–93.

Milbourne, P. (2011) 'Poverty in rural wales: material hardship and social inclusion', in P. Milbourne (ed) *Rural Wales in the Twenty-First Century: Society, Economy and Environment*, Cardiff: University of Wales, pp 254–72.

Milbourne, P. (2014) 'Poverty, place and rurality: material and sociocultural disconnections', *Environment and Planning A*, 46: 566–80.

Milbourne, P. (2016a) 'Austerity, welfare reform and older people in rural places: competing discourses of voluntarism and community', in N. Hanlon and M. Skinner (eds) *Ageing Resource Communities: New frontiers of Rural Population Change, Community Development and Voluntarism*, New York: Routledge, pp 74–88.

Milbourne, P. (2016b) 'Poverty and welfare in rural places', in M. Shucksmith and D.L. Brown (eds) *Routledge International Handbook of Rural Studies*, London: Routledge, pp 450–61.

Milbourne, P. and Coulson, H. (2020) 'Tackling rural poverty in Wales: developing community-based approaches. A position paper for the Rural Futures project', Cardiff: Cardiff University.

Ministry of Housing, Communities & Local Government (2019) 'English indices of deprivation', London: UK government.

Mirowski, P. (2013) *Never Let a Serious Crisis Go to Waste*, London: Verso.

Moffatt, S. and Scrambler, G. (2008) 'Can welfare-rights advice targeted at older people reduce social exclusion?' *Ageing and Society*, 28: 875–99.

Moseley, M. (2003) *Local Partnerships for Rural Development: The European Experience*, Wallingford: CABI Publishing.

National Audit Office (2002) 'Tackling pensioner poverty: encouraging take-up of entitlements', report by the Comptroller and Auditor General, House of Commons 37, Session 2002–03, 20 November.

National Audit Office (2018) 'Rolling out Universal Credit', report by the Comptroller and Auditor General, House of Commons 1123, Session 2017–19, 15 June.

National Audit Office (2021) 'Local government finance in the pandemic', report by the Comptroller and Auditor General, House of Commons 1240, Session 2019–21, 10 March.

Newby, H. (1980) *Green and Pleasant Land? Social Change in Rural England*, Harmondsworth: Penguin.

Northumberland County Council (2019) 'Indices of deprivation 2019 – Northumberland summary report', Morpeth: Northumberland County Council.

Northumberland County Council (2020) 'Northumberland knowledge economic performance bulletin', Morpeth: Northumberland County Council.

Northumberland National Park (2020) 'Local plan', Hexham: Northumberland National Park.

OECD (2020) 'Policy implications of coronavirus crisis for rural development', Paris: OECD.

Office for National Statistics (ONS) (2016) '2011 urban rural classification', London: ONS.

Office for National Statistics (ONS) (2018) 'Population estimates for parishes in England and Wales, mid-2002 to mid-2017', London: ONS.

Outer Hebrides Community Planning Partnership (2017) 'How good is our place?', Stornoway: OHCPP.

Outer Hebrides Community Planning Partnership (2019) 'Anti-poverty strategy 2019–2024', Stornoway: OHCPP.

Pacione, M. (2004) 'The geography of disadvantage in rural Scotland', *Tijdschrift voor Economische en Sociale Geografie*, 95(4): 375–91.

Painter, J. and Philo, C. (1995) 'Spaces of citizenship: an introduction', *Political Geography*, 14: 107–20.

Palmer, G. (2009) 'The poverty site' [online], available from: www.poverty.org.uk/ [accessed 15 January 2022].

Peck, J. (2012) 'Austerity urbanism', *City*, 16(6): 626–55.

Peck, J. and Tickell, A. (2002) 'Neoliberalizing space', *Antipode*, 34(3): 380–404.

Perth and Kinross Council (2015a) 'East Perthshire locality profile', Perth: Perth and Kinross Council.

Perth and Kinross Council (2015b) 'Place-based scrutiny East Perthshire, January–April 2015', Perth: Perth and Kinross Community Planning Partnership.

Perth and Kinross Council (2020) 'Economic impacts of Covid', presentation given by the Head of Planning, Perth and Kinross Council.

Philip, L. and Shucksmith, M. (2003) 'Conceptualising social exclusion in rural Britain', *European Planning Studies*, 11(4): 461–80.

Phillipson, J., Gorton, M., Turner, R., Shucksmith, M., Aitken-McDermott, K., Areal, F., et al (2020) 'The COVID-19 pandemic and its implications for rural economies', *Sustainability*, 12(10): 3973.

Phimister, E., Shucksmith, M. and Vera-Toscano, E. (2000) 'The dynamics of low pay in rural households: exploratory analysis using the British household panel survey', *Journal of Agricultural Economics*, 51(1): 61–76.

Plunkett Foundation (2021) 'Unlocking the potential of community business to boost the rural economy', Woodstock: Plunkett Foundation.

Powell, J., Keech, D., Reed, M. and Dwyer, J. (2018) 'What works in tackling rural poverty', Cardiff: Wales Centre for Public Policy.

Public Policy Institute for Wales (2016) 'Rural poverty in Wales: existing research and evidence gaps', Cardiff: PPIW.

Pugh, R., Scharf, T., Williams, C. and Roberts, D. (2007) 'Obstacles to using and providing rural social care', Social Care Institute for Excellence research briefing 22.

Rank, M. and Hirschl, T. (1993) 'The link between welfare participation and population density', *Demography*, 30: 607–22.

Rank, M., Hirschl, T. and Foster, K. (2014) *Chasing the American Dream: Understanding What Shapes Our Fortunes*, Oxford: Oxford University Press.

Ray, C., Sissons, P., Jones, K. and Vegeris, S. (2014) 'Employment, pay and poverty', report for the Joseph Rowntree Foundation, London: The Work Foundation.

Reimer, W. (1998) 'Personal communication with Professor Mark Shucksmith', as discussed in Philip, L. and Shucksmith, M. (2003) 'Conceptualising social exclusion in rural Britain', *European Planning Studies*, 11(4): 461–80.

Riddoch, L. (2020) *Huts: A Place Beyond – How to End Our Exile from Nature*, Edinburgh: Luath Press.

Robinson, C. and Mattioli, G. (2020) 'Double energy vulnerability: spatial intersections of domestic and transport energy poverty in England', *Journal of Energy Research and Social Science*, 70: 101699.

Room, G. (1994) *Beyond the Threshold*, Bristol: Policy Press.

Ross, D. (2020) 'Built-in resilience: community landowners' responses to the COVID-19 crisis', Community Land Scotland, Greenock: report for Community Land Scotland and the Community Woodlands Association.

Rural Coalition (2022) 'Rural coalition response to the Levelling Up White Paper' [online], March, available from: https://acre.org.uk/rural-coalit ion-response-to-the-levelling-up-white-paper/ [accessed 20 May 2022].

Rural England CIC (2022) 'The state of rural services 2021: the impact of the pandemic', Rural England CIC, Craven Arms: Rural England CIC.

Rural Services Network (2014) 'Warning over rural childcare costs', *Rural Services Network* [online], 13 May, available from: www.rsnonline.org.uk/ services/warning-over-rural-childcare-costs [accessed 15 January 2022].

Russell, M. (2020) 'Plans to make Harris a "control area" for holiday lets', *West Highland Free Press* [online], 17 October, available from: www.whfp. com/2020/10/17/plans-to-make-harris-a-control-area-for-holiday-lets/ [accessed 15 January 2022].

Satsangi, M. and Wilson, M. (2020) 'Do housing costs impact on poverty in rural areas?', *Housing Studies*, 35(3): 415–38.

Savage, M. (2010) *Identities and Social Change in Britain since 1940: The Politics of Method*, Oxford: Oxford University Press.

Savage, M., Bagnall, G. and Longhurst, B. (2005) 'Local habitus and working class culture', in Devine, F., Savage, M., Scott, J. and Crompton, R. (eds) *Rethinking Class: Culture, Identity and Lifestyles*, Basingstoke: Palgrave Macmillan, pp 95–122.

Scott, A., Gilbert, A. and Gelan, A. (2007) 'The urban–rural divide: myth or reality?', Socio-Economic Research Group Policy Brief, Aberdeen: The Macaulay Institute.

Scott, C., Sutherland, J. and Taylor, A. (2018) 'Affordability of the UK's Eatwell Guide', London: The Food Foundation.

Scottish Government (2018a) 'Understanding the Scottish rural economy', Edinburgh: Scottish Government.

Scottish Government (2018b) 'Rural Scotland key facts 2018', Edinburgh: Scottish Government.

Scottish Government (2019) 'Local level Brexit vulnerabilities in Scotland: Brexit Vulnerabilities Index', Edinburgh: Scottish Government.

Scottish Government (2021) 'Social security', *Scottish Government* [online], available from: www.gov.scot/policies/social-security/ [accessed 15 January 2022].

Scottish Government (2021a) 'Poverty in rural Scotland: a review of evidence', Edinburgh: Scottish Government.

Scottish Government (2021b) 'The cost of remoteness – reflecting higher living costs in remote rural Scotland when measuring fuel poverty', Edinburgh: Scottish Government.

Scottish Household Survey (2012) 'Scottish household survey 2018', Edinburgh: Scottish Government.

Shaw, J.M. (1979) *Rural Deprivation and Planning*, Norwich: Geo Books.

Sherman, J. (2006) 'Coping with rural poverty: economic survival and moral capital in rural America', *Social Forces*, 85: 891–913.

Sherman, J. (2009) *Those Who Work, Those Who Don't: Poverty, Morality and Family in Rural America*, Minneapolis: University of Minnesota Press.

Sherman, J. (2013) 'Surviving the great recession: growing need and stigmatized safety net', *Social Problems*, 60(4): 409–32.

Sherman, J. (2021) *Dividing Paradise: Rural Inequality and the Diminishing American Dream*, Oakland: University of California Press.

Shortall, S. and Alston, M. (2016) 'To rural proof or not to rural proof: a comparative analysis', *Politics & Policy*, 44(2): 35–55.

Shucksmith, M. (2000) 'Exclusive countryside? Social inclusion and regeneration in rural areas', York: Joseph Rowntree Foundation.

Shucksmith, M. (2001) 'History meets biography: processes of change and social exclusion in rural areas', presentation at the 'Exclusion Zones: Inadequate Resources and Civic Rights in Rural Areas' Conference, Queen's University, Belfast.

Shucksmith, M. (2004) 'Young people and social exclusion in rural areas', *Sociologia Ruralis*, 44(1): 43–59.

Shucksmith, M. (2008) 'New Labour's rural policy in international perspective', in M. Woods (ed) *New Labour's Countryside*, Bristol: Policy Press, pp 59–78.

Shucksmith, M. (2010) 'Disintegrated rural development: neo-endogenous rural development, planning and place-shaping in diffused power contexts', *Sociologia Ruralis*, 50(1): 1–14.

Shucksmith, M. (2012) 'Class, power and injustice in rural areas: beyond social exclusion?', *Sociologia Ruralis*, 52(4): 377–97.

Shucksmith, M. (2016) 'Rural deprivation – national trends and implications for rural communities', presentation to the Community Action Northumberland Spring Conference.

Shucksmith, M. (2018a) 'Social exclusion in rural places', in M. Shucksmith and D. Brown (eds) *Routledge International Handbook of Rural Studies*, London: Routledge, pp 433–49.

Shucksmith, M. (2018b) 'Reimagining the rural: from rural idyll to good countryside', *Journal of Rural Studies*, 59: 163–72.

Shucksmith, M. (2019) 'Rural policy after Brexit', *Contemporary Social Science*, 14(2): 312–26.

Shucksmith, M., Chapman, P. and Clark, G. (1994) 'Disadvantage in rural Scotland: how is it experienced and how can it be tackled?', York: Joseph Rowntree Foundation.

Shucksmith, M., Chapman, P. and Clark, G. (1996) *Rural Scotland Today: The Best of Both Worlds?*, Aldershot: Avebury.

Shucksmith, M. and Chapman, P. (1998) 'Rural development and social exclusion', *Sociologia Ruralis*, 38(2): 225–42.

Shucksmith, M. and Schafft, K. (2012) 'Rural poverty and social exclusion in the United States and the United Kingdom', in M. Shucksmith, D. Brown, S. Shortall, J. Vergunst and M. Warner (eds) *Rural Transformations and Rural Policies in the US and UK*, New York: Routledge, pp 100–16.

Shucksmith, M., Davoudi, S., Todd, L. and Steer, M. (2021) 'Conclusion: hope in an age of austerity and a time of anxiety', in M. Steer, S. Davoudi, M. Shucksmith and L. Todd (eds) *Hope under Neoliberal Austerity: Responses from Civil Society and Civic Universities*, Bristol: Policy Press, pp 257–74.

Simcock, N., Jenkins, K., Mattioli, G., Lacey-Barnacle, M., Bouzarfovski, S. and Martiskainen, M. (2020) 'Vulnerability to fuel and transport poverty', CREDS Policy brief 010, Oxford: Centre for Research into Energy Demand Solutions.

Sindico, F., Sajeva, G., Sharman, N., Berlouis, P. and Ellsmoor, J. (2020) 'Islands and COVID-19: a Global Survey', Strathclyde: University of Strathclyde.

Skelton, L.J. (2014) 'The uncomfortable path from forestry to tourism in Kielder, Northumberland: a socially dichotomous village?', *Oral History*, 42(2): 81–93.

Skerratt, S. and Woolvin, M. (2014) 'Rural poverty and disadvantage: falling between the cracks?', Rural Scotland in Focus 2014 report, Edinburgh: Scotland's Rural College.

Smith, N., Davis, A. and Hirsch, D. (2010) 'A minimum income standard for rural households', Cheltenham: Joseph Rowntree Foundation and Commission for Rural Communities.

Social Enterprise in Scotland (2019) 'Census 2019', *Social Value Lab* [online], available from: https://socialenterprisecensus.org.uk/ [accessed 15 January 2022].

Social Mobility Commission (2019) 'State of the nation 2018–19: social mobility in Great Britain', London: UK Government.

Standing, G. (2011) *The Precariat: The New Dangerous Class*, London: Bloomsbury.

Stoker, G. (1998) 'Governance as theory: five propositions', *International Social Science Journal*, 155: 17–28.

The Courant (2020) 'National park housing boost', 23 July.

Thomson, S., McMorran, R., Bird, J., Atterton, J., Pate, L., Meador, E., de Lima, P. and Milbourne, P. (2018) 'Farm workers in Scottish agriculture: case studies in the international seasonal labour market', Edinburgh: Scottish Government.

Tickamyer, A. and Henderson, D. (2011) 'Livelihood practices in the shadow of welfare reform', in K.E. Smith and A.R. Tickamyer (eds) *Economic Restructuring and Family Well-Being in Rural America*, University Park: Penn State Press, pp 294–319.

Tickamyer, A., Sherman, J. and Warlick, J. (2017) *Rural Poverty in the United States*, New York: Columbia University Press.

Tighean Innse Gall (2022) 'Statement from Tighean Innse Gall (TIG) – 2nd March 2022' [online], available from: https://tighean.co.uk/statement-from-tighean-innse-gall-tig-2nd-of-march-2022/ [accessed 20 May 2022].

Tonts, M. and Larsen, A.C. (2002) 'Rural disadvantage in Australia: a human rights perspective,' *Geography*, 87: 132–41.

Townsend, P. (1979) *Poverty in the United Kingdom*, Harmondsworth: Penguin.

Townsend, P. (1987) 'Disadvantage', *Journal of Social Policy*, 16: 125–46.

Vera-Toscano, E., Shucksmith, M. and Brown, D. (2020) 'Poverty dynamics in rural Britain 1991–2008: did Labour's social policy reforms make a difference?', *Journal of Rural Studies*, 75: 216–28.

Vera-Toscano, E., Shucksmith, M., Brown, D. and Brown, H. (forthcoming) 'West Northumberland Foodbank (2021) 'West Northumberland food bank project report year 7', Hexham: West Northumberland Foodbank.

Westholm, E., Moseley, M. and Stenlas, N. (1999) 'Local partnership and rural development in Europe: a literature review of practice and theory', Falun, Sweden: Dalarna Research Institute report.

Willet, J. (2021) *Affective Assemblages and Local Economies*, London: Rowman and Littlefield.

Williams, R. and Doyle, R. (2016) 'Rural poverty in Wales', Cardiff: Public Policy Institute for Wales.

Wilson, B. (2019) 'State of rural services 2018', Craven Arms: Rural England.

Wilson, R. and Copus, A. (2018) 'Services of General Interest (SGI) in the Scottish Sparsely Populated Area (SPA): introduction, classification by delivery mode, and selection of exemplar services', Edinburgh: Scottish Government.

Women's Resource Centre (nd) 'CEDAW article 14: a case study of issues affecting rural women in the north east of England', London: Women's Resource Centre.

Woods, M. (2006) *Rural Geography, Processes, Responses and Experiences in Rural Restructuring*, London: Sage Publications.

Woods, M. (2011) *Rural*, London: Routledge.

Woods, M., Goodwin-Hawkins, B. and Jones, R. (2019) 'The right to the country? Ruralising spatial justice', abstract of paper presented at the 'Rural Futures in a Complex World' conference, Trondheim, Norway, June.

Woodward, R. (1996) '"Deprivation" and the "rural": an investigation into contradictory discourses', *Journal of Rural Studies*, 12: 55–67.

Wyn, J. and White, R. (2015) 'Complex worlds, complex identities: complexity in youth studies', in D. Woodman and A. Bennett (eds) *Youth Cultures, Transitions and Generations: Bridging the Gap in Youth Research*, Basingstoke: Palgrave Macmillan, pp 28–41.

Yarwood, R. (2017) 'Rural citizenship', in D. Richardson, N. Castree, M.F. Goodchild, A. Kobayashi, W. Liu and R.A. Marston (eds) *International Encyclopedia of Geography: People, the Earth, Environment and Technology*, Hoboken: Wiley-Blackwell.

Index

.